OSPREY AIRCRAFT OF THE ACES • 2

Bf 109 Aces of North Africa and the Mediterranean

SERIES EDITOR: TONY HOLMES

OSPREY AIRCRAFT OF THE ACES · 2

Bf 109 Aces of North Africa and the Mediterranean

Jerry Scutts

OSPREY
AEROSPACE

First published in Great Britain in Autumn 1995
by Osprey Publishing. Michelin House, 81 Fulham Road,
London SW3 6RB

Reprinted Summer 1995
Second Reprint Spring 1996, Summer 1998

© 1994 Osprey Publishing Limited
© 1994 Osprey Publishing/Aerospace Publishing colour side-views

All rights reserved. Apart from any fair dealing for the purpose of private study,
research, criticism or review, as permitted under the Copyright, Design and
Patents Act, 1988, no part of this publication may be reproduced, stored in a
retrieval system, or transmitted in any form or by any means, electronic, electrical,
chemical, mechanical, optical, photocopying, recording or otherwise, without prior
written permission. All enquiries should be addressed to the publisher.

ISBN 1 855324482

Edited by Tony Holmes
Design by TT Designs, Tony & Stuart Truscott

Cover Artwork by Iain Wyllie
Aircraft Profiles by Chris Davey, John Weal and Iain Wyllie
Figure Artwork by Mike Chappell
Scale Drawings by Mark Styling

Printed in Hong Kong

Acknowledgements:
Osprey duly acknowledge the published works of Christopher Shores and
co-authors (**Fighters over the Desert, Fighters over Tunisia and Malta** and
The Hurricane and Spitfire Years) and Eduard Neumann in the preparation of
this manuscript.

Editor's Note
To make this new series as authoritative as possible, the editor would be extremely
interested in hearing from any individual who may have relevant photographs,
documentation or first-hand experiences relating to the elite pilots, and their
aircraft, of the various theatres of war. Any material used will be fully credited to its
original source. Please write to Tony Holmes at 1 Bradbourne Road, Sevenoaks,
Kent, TN13 3PZ, Great Britain.

Front cover
By the Spring of 1942 the 'star' of JG
27 in North Africa, Hans-Joachim
Marseille, was shinning at its
brightest. The seamless combination
of the *Experten* and his weapon of
war – the Bf 109F – proved more
than a match for the RAF in-theatre,
and Marseille duly cut swathes
through the Allied ranks of
Kittyhawks and Hurricanes as he
raced towards the 100 victories mark
(Cover painting by Iain Wyllie)

CONTENTS

DESERT JAGDFLIEGER

Having declared war on Britain and France on 11 June 1940, Benito Mussolini began an immediate air offensive to knock out the island of Malta, a British base vital to any plans to wage war in the Middle East. Its location at the 'crossroads' of the Mediterranean sea lanes meant that whoever held the island could command the north-south route from Italy to North Africa, and the east-west route from Gibraltar to Alexandria and the Suez Canal.

Italy's aggression was not entirely unexpected and the early raids were countered by a handful of Sea Gladiators, the RAF giving the attackers as hot a reception as possible. Over the ensuing seven months through to the New Year of 1940, the Regia Aeronautica failed to destroy Malta's defences, although it made a determined effort to do so. The first Hurricanes reached the beleaguered island in June, and by December the Luftwaffe had arrived in Sicily to assist the Italians. *X.Fliegerkorps* made its headquarters at Catania, with Palermo, Trapani, Gela and Comiso its main airfields on the island.

Initally, the German's aerial strength was confined to Stuka and bomber units, with Bf 110s in support, but in February 1941 14 Messerschmitt Bf 109Es from 7./JG 26 flew to Sicily to open a new phase of the air war in the Mediterranean. Beginning escort sorties over the island almost immediately, the JG 26 detachment began an unparalleled run of good fortune. Sweeping aside the weak RAF defences, the German pilots strafed Maltese airfields at will, shot down defending Hurricanes almost with impunity, and generally made life very unpleasant for the defenders, who were already having to put up with a constant pounding from Italian and German bombs.

Although the RAF put up numerous interception sorties and succeeded in ensuring that the bombing was rarely achieved without some cost to the attackers, it was singly unable to counter the Bf 109s effectively. Many of the Allied pilots had had relatively little combat time, and their Hurricanes were in the main well-used. Claims were filed by a number of pilots who swore that they had most definitely shot down a Bf 109, but subsequent scrutiny of Luftwaffe records showed that the *Jagdflieger's* survival rate over Malta at this stage of the

The 'advance guard' of a more extensive involvement in the Western Desert by the *Jagdwaffe* was the handful of Bf 109E-7s of 7. *Staffel*, JG 26 *Schlateger*. Led by *experten* Oblt Joachim Müncheberg, the *Staffel's* 14 aircraft arrived at the Sicilian airfield of Gela on 9 February 1941, primarily to support the Regia Aeronautica's assault on Malta. Here, following yet another successful Hurricane destroying sortie over the island, Müncheberg climbs out of his personal Bf 109E-7 to be greeted by one of his groundcrewman (at right), who is clutching a traditional aces' laurel wreath to mark the occasion. Note the Oberleutnant's white metal rank pennant on the radio mast of 'White 12'

Also part of 7./JG 26 was Fw Karl Laub, a seasoned NCO pilot of the *Staffel* who scored seven kills up to his death on 14 December 1944. Note the wheel covers which were used to stop tyre rubber perishing in the extreme Mediterranean heat. Laub's E-7 displays both the distinctive red heart *Staffel* badge on its yellow nose, plus the *Schlageter Geschwader* 'S' just forward of the cockpit

En route to Sicily, 7./JG 26 stopped briefly in the Balkans, where an anonymous *Jagdflieger* tried his hand at a less powerful form of transport than that which he was normally accustomed to. Taking no chances with an unfamiliar type, the young pilot seems well-padded against an unexpected ground loop! The 'Mk I Donkey's' normal 'pilot' appears somewhat bemused by the whole episode

war was nothing short of miraculous. Few Bf 109s were damaged, let alone shot down, and Oblt Joachim Müncheberg's small force dictated the progress of local air combat on more or less their own terms.

Fortunately for the defenders of Malta, Hitler decided to invade the Balkans on 6 April, thus drawing off most of the fighter strength in Sicily, including the modest 7./JG 26 force, which moved to bases in Italy to put it within range of targets in Yugoslavia. With the Luftwaffe also committed in Greece, Malta enjoyed a brief respite, at least from incursions by the *Schlageter Geschwader*'s Messerschmitts. On 9 April, however, they returned to Sicily.

Flying a typical high altitude sweep around the island on 27 April, Müncheberg, and his able wingman Oblt Mietusch, spotted a Sunderland approaching its anchorage. The flying boat had been leading in a flight of Hurricanes that had earlier launched from the carrier HMS *Ark Royal*, and just as it was being made fast, the Bf 109s swept in. A single pass was enough to set the Sunderland ablaze, and it quickly became a total loss. This was the second Short flying boat destroyed by JG 26, a single aircraft having been sunk at its moorings the month before – another Sunderland had also been damaged in the attack.

Müncheberg, himself, had done well over Malta, scoring his 39th and 40th victories on 1 May. 7./JG 26's presence had also severely shaken the defenders' morale, although the detachment's days in the theatre were numbered. On 25 May Müncheberg showed how complete his mastery of the air was by leading seven Bf 109s on a low-level strafing attack on Ta Kali airfield, which was one of the main RAF stations on Malta. Sweeping in from the south, a direction not usually taken, the Germans surprised the defenders and left five newly-delivered Hurricane Mk Is of No 249 Sqn on fire following two passes over the airfield. It was a farewell gesture typical of Müncheberg, and he headed back to Sicily, bound for combat in Greece, more than satisfied with his 'tour' of Malta!

Meanwhile, further German help for Mussolini had arrived in the Mediterranean in April when I./JG 27 occupied Ain El Gazala airfield in Libya, and flew its first sortie on the 19th. The presence of German fighters on the mainland was to help Italian ground forces who had been locked in bitter combat with the British in Cyrenacia for some ten months. Up to this point in the campaign things had gone badly for the Axis forces in the Italian colony, and they were in serious danger of being driven out of North Africa all together. Mussolini's longer term plans of bringing all of Egypt under fascist domination appeared to have been thwarted before they had really begun. Under the terms of his 'Pact of Steel' with Italy, Hitler felt obliged to militarily assist the Italian dictator in a war theatre for which he personally had few plans. It would also later transpire that the *Führer* had not been informed of Mussolini's ambitious plans for expansion in Greece and Ethiopia either!

7./JG 26 arrived in the Mediterranean from its Battle of Britain airfield in the Pas de Calais with a fine reputation, carved during the long summer of the previous year. During its brief stay in Sicily, the small unit was to become the scourge of the RAF Hurricane squadrons charged with defending Malta

Photographed at Catania, in Sicily, in early April 1941. *Staffelkapitan*, and *Ritterkreuztrager*, Hptm Karl-Heinz Redlich (left foreground, holding papers) briefs his I./JG 27 pilots before their long overwater flight to Ain El Gazala, in Libya *(Schroer)*

NORTH AFRICA

Under the tactical control of *X.Fliegerkorps*, JG 27's three *Staffeln* were despatched to North Africa with the tough and resourceful Hptm Eduard Neumann as their commander. The *Gruppe* was composed mostly of experienced pilots, many with multiple victories to their credit. Like 'Edu' Neumann himself, individuals such as Oblt Ludwig Franzisket (14 kills), Oblt Karl-Heinz Redlich (10) and Lt Willi Kothmann (7) arrived in the desert thoroughly familiar with their 'Emils', having flown the nimble fighter during both the Blitzkrieg of Western Europe and the Battle of Britain. Neumann's history with the Messerschmitt fighter stretched back even further, as he had flown Bf 109Ds with the *Condor Legion* in Spain, where he opened his own score with the destruction of two Republican aircraft.

Similarly impressive combat records were boasted by other JG 27 pilots, the victory list of the most successful being headed at that time by Oblt Gerhard Homuth with 15. Some way down this roll was Oberfh Hans-Joachim Marseille with 7. The respective I *Gruppe Staffelkapitanen* were Redlich (ten victories) (1./), Hptmn Erich Gerlitz (three victories) (2./) and Homuth (3./). The overall victory tally for I./JG 27 pilots on arrival in the desert stood at 61.

Although it only possessed the standard full *Gruppe* strength of about 90 aircraft, I./ JG 27 posed a very real threat to Allied air operations. In their new arena, the *Jagdflieger* faced RAF units generally flying aircraft that had seen much service, ex-Battle of Britain Hurricanes being the premier Allied fighter throughout the region. Britain had been forced to give the Middle East a low priority until the threatened German cross-Channel assault of 1940 had been contained. A trickle of new and replacement aircraft had hardly begun to arrive in Egypt by the time JG 27 went operational in Libya. Nevertheless, Allied aircraft outnumbered the Germans, a situation that was to prevail even when additional Axis units were sent to North Africa.

Established at Ain El Gazala, located on the Mediterranean coast facing the Gulf of Bomba between

Surrounded by empty fuel drums, and boasting a much loathed centreline drop tank, the tropicalised Bf 109E-7 of I./JG 27 *Gruppen* Adjutant Oblt Ludwig Franzisket is warmed up at Catania prior to departing for North Africa – several dusty and slipstream-beaten groundcrewmen keep a firm hold on the tail whilst their pilot pushes the throttles of the aircraft's, DB601A engine up against the quadrant stops to check the powerplant's operability *(Schroer)*

When I./JG 27 transferred to the Western Desert, a modest ceremony was held to formerly install the *Staffel* in a new theatre of war – they were also presented with a Luftwaffe standard to mark the occasion. A freshly painted Bf 109E-7/Trop, decorated with I.*Gruppe*'s famous 'Leopard over Africa' emblem, forms a suitable backdrop for the event, which was almost certainly held at Ain El Gazala

Tobruk and Tmimi, JG 27 used Gambut as a forward base, positioning one *Staffel* there for the day's sorties, with the fighters returning to Ain El Gazala each evening. In the early days pilots purchased locally personal items of kit from Arab (and Jewish) traders including clothing, as there had been little provisioning made by the Luftwaffe general staff for protracted periods of flying in desert conditions. Pilots did, however, have their own regulation issue uniforms and lighter-weight 'summer' kit, which they wore most of the time, matched with warmer clothing as necessary.

In broad terms the Western Desert had a pleasant enough climate on the coastal strip where most of the airfields were located, although the maxim that high daytime temperatures were invariably followed by extremely cold nights had to be allowed for by anyone intending a protracted stay. The Germans also realised that in regard to provisions and equipment, all of which was initially in short supply, the main North African ports of Tripoli and Benghazi lay at the end of a very long supply line. Aircraft spares were a continual problem and maintenance had to be carried out in the open, hampered by a general lack of tools, heavy lifting equipment and vehicles.

While the Bf 109E was becoming outmoded in other theatres, it was more than adequate for JG 27's task. This became abundantly clear after the first clashes with enemy aircraft; any inherent drawbacks with the 'Emil' were easily compensated for by pilots who already had hundreds of hours on the type, and had a great affection for it.

To prevent excessive engine wear on Bf 109Es operating under sandy or dusty conditions, a number of 'tropicalisation' modifications had been made in 1940. These included a filter for the port side ram air intake for the Daimler Benz DB 601A engine, lubricant cooling and generator ventilation. In addition, pick up points for a 300 lt (66 Imp gal) belly drop tank were also fitted, and despite the *Jagdfliegers'* dislike of these items, due mainly to their inherent fire risk, extra tanks were useful for long flights over the sea. JG 27's aircraft carried drop tanks en route to North Africa, but little use appears to have been made of them subsequently.

SOUND TACTICS

While the *Jagdflieger* were often obliged to operate at low to medium altitudes escorting dive and medium bombers, the Luftwaffe's doctrine of always maintaining a high cover held good. Flying the 'finger four' formation that had proven so effective since the Spanish Civil War, the Bf 109Es usually operated in small numbers, often in no more than *rotte* (two) or *schwarme* (four) strength. An altitude of 6000 ft was generally favoured by the pilots, patrols being mounted from daybreak to around 1000 hrs, and from 1600 to dusk, when the angle of the sun and the ever-present dust haze put the crews of any aircraft flying lower at some disadvantage. These times were also advantageous to dive bomber crews, who could often reach their targets virtually undetected, the defenders only spotting the diving Stakas when it was far too late.

Although acting as a support force for Gen Erwin Rommel's Afrika Korps, JG 27 enjoyed a remarkable degree of flexibility and freedom to plan its own operations. Often sorties – particularly *Frie Jagd* patrols – were only indirectly supportive, the whittling down of the RAF's offensive strength being seen as a more effective employment of the meagre *Jagdflieger* resources as it made the movement of German supplies by road and air that much safer. And it also left the *Gruppe's experten* free to pick off numerous unsuspecting stragglers from enemy formations.

Sweeps and strafing runs took their toll of the RAF, as did the interception of low flying Allied aircraft engaged on similar work against the Germans. Strafing was widely practised by both sides, for such sorties resulted in the loss of personnel, destroyed or badly damaged aircraft and a reduction in fuel supplies and equipment. By using nimble fighters for such missions, results were achieved at a lower cost in terms of both men and machines than might have been the case with slower bombers.

The nature of the desert air war in 1941 was such that the *Jagdflieger* most frequently met enemy fighters in combat, rather than bombers or reconnaissance aircraft, and unsurprisingly these were to represent the bulk of their aerial victories. The latter types were very much in short supply during the early months of the campaign, although a small number of kills against Blenheims, Beauforts and Marylands were achieved.

Due to a common reliance on seaborne supplies, the main fighting in the Western Desert was confined to the coastal areas of Libya, a series of battles for the towns with seaports, and their nearby airfields, raging throughout the three-year conflict. The single main coast road was a lifeline for both sides, and possession of major sections of it was the key to timely reinforcements of armour, vehicles and personnel.

Above all, it was equally vital for both the Axis and Allied forces that they maintain fuel and oil supplies if the opposing armies were to continue the fight, and it is somewhat ironic, considering the subsequent history of this region, that the lubricants of war were not obtained locally, but in the Luftwaffe's case shipped to Tripoli by tanker from Germany. Stored in drums, fuel was stockpiled at airfields and hidden under camouflage netting, or preferably stored underground, which both hid it from prying enemy reconnaisance and also kept it cool. Subterranean quarters were also excavated for personnel for much the same reason.

The Luftwaffe cameraman despatched to Libya with I./JG 27 made the most of this inauguration ceremony by recording the appearance of Bf 109E-7 Trop 'Yellow 4' from several angles. His photographic essay is invaluable all these decades later as it shows the very first desert camouflage scheme applied by the *Staffel* to their 21 Bf 109s. The only weathering of the paint visible in this shot is the staining of the fuselage over the wing produced by the engine exhaust stubs. Photographed only days after arriving in Libya, this Messerschmitt had seen little action up to this point as the unit's pilots were still familiarising themselves with their new combat environment

TOBRUK

By April 1941 the Afrika Korps had fought its way to Tobruk and had the town under virtual seige. German fighters based at Ain El Gazala, which was only eight miles west of the port town, gave oustanding protection to the Ju 87B-2s of II./StG 2, which repeatedly bombed the defenders. Allied counter attacks took the form of bombing and ground strafing

German positions, and opposing fighter formations frequently clashed Many German pilots were amazed at the tactics of the RAF and South African Air Force (SAAF) squadrons. In comparison with their own well rehearsed and flexible 'finger four' formations, the British appeared to some to fly in unwieldy flocks, 'like a bunch of grapes', as one pilot described them. Whenever they had the chance Allied pilots would form a defensive circle, using the manoeuvrability of the Hurricane to turn tightly. *Jagdflieger* who possessed great nerve delighted in risking their lives by penetrating the circle and picking off individual targets. Others overcame their fear of being put at a disadvantage by what amounted to a ring of enemy guns, and emulated pilots of the calibre of Marseille, who delighted in flying such tactics.

The core around which German fighter tactics were built was speed. Dive, shoot, zoom away and avoid a dogfight were maxims that rarely needed to be varied. They were adopted wherever the *Jagdwaffe* flew, and were extremely effective over the desert. The clarity of the desert weather also aided the alert pilot. Allied aircraft were frequently spotted well before they had a chance to react, and victories could be confirmed with ease as burning or intact wrecks would be observed at great distances in the featureless terrain. *Jagdflieger* would often drive out to inspect their victims at close quarters and obtain souvenirs in the time-honoured tradition.

In such an atmosphere the Germans thought nothing of taking on formations many times their own size. Superior tactics often saw them emerge as the victors, but there were rarely enough Bf 109s on hand to destroy whole formations of fighters or bombers piecemeal, and it was clear that while this prevailed, little gain would be achieved by either side.

Much the same could also be said for the ground forces, where a general parity in tanks prevented any decisive engagements taking place. But the secret of the Allies being able to hold on in the face of Rommel's bold thrusts was a gradual building of superb air-to-ground co-operation, which the Germans never achieved. On the other hand, the Luftwaffe enjoyed the help of its Italian partners and it had, unlike the RAF, a strong air tranport force which immeasurably boosted the flow of supplies. While most Germans got on well with their Italian counterparts, there was widespread feeling that poor tactics robbed the latter of many opportunities to exploit the qualities of their aircraft. This was particularly true when the Regia Aeronautica introduced the Macchi C.202 in the autumn of 1941.

When Rommel temporarily abandoned his attempt to capture Tobruk, the ground war developed into a series of attacks and counterattacks, the Allies taking the offensive on a limited scale in a number of operations. Two such attacks, codenamed *Brevity* and *Crusader*,

One of the Luftwaffe's most able and long-serving officers, Hptm Eduard 'Edu' Neumann led I./JG 27 during its early, halcyon, days. He is pictured here in 'typical' desert uniform – comfort, rather than regulation, was the watchword in Libya for the *Jagdflieger* when it came to choosing flying clothing

Most of JG 27's servicing was carried out in the open, so a dust filter over the engine air intake was vital if the fine, abrasive, sand grains ingested during ground taxying and take-off were not to wreck the DB 601 – sand remained a constant problem throughout the campaign

A *rotte* of Bf 109E-7/Trops over the barren terrain of Libya. If a pilot was forced to bail out during a sortie he could quickly die in the heat of the desert, so JG 27 maintained a number of Fieseler Storch aircraft to quickly rescue downed *Jagdflieger*

As the location of desert airfields was well known to the enemy, the Germans quickly adapted to local conditions to render their aircraft as inconspicuous as possible. This Bf 109E-7/Trop shows the highly effective green dapple scheme worn by most Messerschmitts – it wasn't so good over water, however!

The badge of I./JG 27, although appropriate to Libyan operations, was actually adopted months before its Bf 109s ever moved to North Africa. It supposedly reflected the 'colonial' outlook then espoused by the Third Reich, and it is believed that the choice of continent was purely coincidental. The face was that of a negress, and such figures were later to figure in *Staffel* lore

achieved very little, but they signalled the start of a campaign of small offensives to contain the Afrika Korps. A decisive result had to await the arrival of convoys to the Mediterranean with fresh troops and tanks.

Just how well the JG 27 pilots knew their trade was reflected in fresh victories almost as soon as they arrived in Libya, but the clock was moving towards the start of a more significant campaign for the Germans. With the successful completion of fighting in Greece, II./JG 27 returned to Germany while III *Gruppe* moved to Sicily on 5 May.

CRETE

The expanding war in the Mediterranean obliged Hitler to also secure Crete before moving the bulk of his forces to what was soon to become the Eastern Front. II. and III./JG 77 and I.(J)/LG 2, with a combined total of 119 Bf 109s, covered the airborne assault under the command of *VIII.Fliegerkorps*. The attack began on 14 May, and the fighters' part remained modest in terms of victories and losses. Once the RAF fighters were either destroyed or evacuated, the Bf 109s set about making the landing areas as safe as possible for the *Fallschirmjäger*, who were ordered to secure the island. Many strafing sorties were flown, and losses to ground fire were heavier than during the brief aerial war, as British troops had fortified many areas and were well dug in. Having called upon III./JG 52 for help, JG 77 actually found that the island was all but secure before the former arrived.

It was in the warm waters that surround Crete that the Bf 109E-4 *Jabos* achieved their greatest successes of the war against the Royal Navy, JG 77 seriously damaging both the battleship HMS *Warspite* and sinking the cruiser HMS *Fiji* on 22 May. *Lehrgeschwader* 2's pilots also did well, downing three Hurricane Mk Is intercepted en route to Crete from Egypt during the remaining days of enemy resistence. Crete, at enormous cost in parachute troops, was totally in German hands by early June.

JG 77 now withdrew to Eastern Europe, as did JG 52, and in Libya JG 27 was grateful for a temporary slow-down in operations after a hectic April. Both 7./JG 26 and I./JG 27 enjoyed a spell of routine patrols, although pilots noted that by June new fighter types had appeared as US Lend-Lease help to the RAF worked through to units in North Africa. Curtiss Tomahawks (P-40B/C) and Kittyhawks (P-40D/E) were at the vanguard of this Allied co-operation, and these were to figure significantly during the coming months. They were flown by RAAF, RAF and SAAF units, but not always effectively from an Allied viewpoint.

The debut of No 2 Sqn SAAF with the Kittyhawk I (along with No 250 Sqn RAF) on 15 July prompted urgent signals to Berlin to have JG 27 re-equipped with the Bf 109F, as it appeared that this new type might be a tougher opponent than the Tomahawk – this was particularly true in regards to its armament, which saw the latter's .303 in guns replaced with six .50 in MGs. Events were to prove otherwise, however.

On its first outing with Kittyhawks, No 2 Sqn's aircraft found a Ju 87 formation. Whilst concentrating on the dive bombers, the Kittyhawk pilots were attacked by eight Bf 109s. Although none of the 14 Axis aircraft were shot down, simply manoeuvring with the South Africans showed the *Jagdflieger* that they would not have much to fear from the new Curtiss fighter – the SAAF came to the same conclusion.

The Luftwaffe rather succinctly described the Kittyhawk as 'Not a very good aircraft to hunt Messserschmitts in!' Provided that the enemy kept using his 'flock formations', it would present them with few problems. SAAF commanders had at least addressed the challenge of the *Jagdfliegers'* intelligent use of height and sun by the time the Kittyhawks arrived in North Africa, duly advising their pilots to fly to altitudes up to 18,000 ft. However, this just resulted in the Germans climbing to altitude soon after take-off, rather than prior to engaging the Allies. The latter appeared to overlook the fact that few aircraft on either side could match the Bf 109's excellent ceiling (up to 36,000 ft), and for a time at least, all Allied fighter types continued to come of worse in air combat.

Müncheberg destroyed two Tomahawks on 29 July before his *Staffel* received orders to return to France. When 7./JG 26 departed in August 1941 its tally of victories since its arrival in Sicily had risen to 52 kills. This score was even more significant as it had been achieved without a single Bf 109 being lost to the RAF.

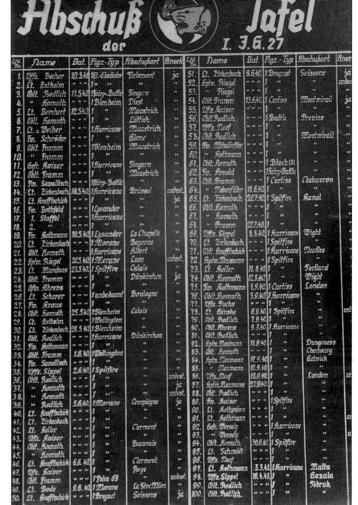

The first 100 victories (*Abschuss*) claimed by I./JG 27 from 10 May 1940 through to 19 April 1941 were accurately recorded on the *Gruppe's* elaborate scoreboard. The unit's first kill in the Mediterranean was actually their 97th of the war, and it fell to Lt Willi Kothmann, who shot down a Hurrican Mk I during a sortie over Malta on 5 March – this was the Leutnant's seventh victory of the war *(J Weal)*

A pristine Bf 109E-7/Trop in Sicily shows the yellow spinner and nose marking (applied over Western European camouflage) that was applied to the first Emils flown by I.*Gruppe* in the Mediterranean. The pilot is holding a belt of distress cartridges that he will soon wrap around the top of his boots *(J Weal)*

'STAR OF AFRIKA'

By the autumn of 1941 the desert armies were still experiencing a lull in the fighting, and there was time enough for I./JG 27 to be withdrawn to Germany *Staffel* by *Staffel* to convert from the Bf 109E-7/Trop to the F-2/Trop. While this was taking place, II.*Gruppe* returned to the Mediterranean on 24 September, fresh from combat in Russia.

Having scored 39 victories in ten days during the early fighting in the East, its leading *experten* was, quite appropriately, the *Gruppenkommandeur*, Hptm Wolfgang Lippert. With 25 kills, he led Oblt Gustav Rödel on 20, Obfw Erwin Sawallisch with 19 and Obfw Otto Shulz, who had scored 9 victories.

At that time the Luftwaffe's tactical strength was deemed adequate to deal with any major British advance, Ain El Gazala and Gambut housing fighters, Tmimi boasting Ju 87B-2s of II./StG 2 and Derna III./ZG 26's Bf 110Ds. In addition, Martuba and Benghazi were alternate airfields used on a rotational basis in line with the Luftwaffe's standard practice of dispersing flying units on a number of adjacent bases. Any one airfield could be utilised without difficulty by fighters, provided that the right grade of fuel was available, and most of the main North African bases consisted of at least two (north and south) landing grounds, if not more.

As in other theatres, squadrons had to take due regard of the weather conditions, and while these were generally favourable for flying, there were a few surprises in store for both sides. Sand storms were expected and compensated for as much as possible, but torrential rain could quickly waterlog airfields and keep aircraft grounded for days on end whilst sodden desert 'runways' soaked up the water that collected on the surface.

On the morning of 3 October 1941 II./JG 27 flew its first sortie since its arrival from Russia, and claimed three Hurricanes. In an afternoon engagement the *Gruppe* swooped down on Sidi Barrani to strafe a dozen Kittyhawks of No 2 Sqn SAAF which had just completed refuelling – the squadron hastily scrambled. Diving too fast, the Germans overshot their targets and Lt D Lacey snapped off a burst of gunfire at the rapidly departing Bf 109s whilst the undercarriage of his Kittyhawk was still retracting.

This lucky shot killed one of the

The majority of JG 27's Emils were repainted in more appropriate 'sand' camouflage with, usually, a white fuselage theatre band. 'White 3', in the foreground, also has a I.*Gruppe* vertical bar just forward of the band. The back and forth nature of the desert war saw both sides abandoning repairable aircraft to the enemy, the Germans, particularly, suffering from a lack of vehicles large enough to move aircraft, often at short notice. Gutted of any salvageable parts, these E-7/Trops were discovered by the Allies neatly lined up at Ain El Gazala in late 1941. Weeks later, they were back in German hands as the airfield fell to the Afrika Korps once again!

Local camouflage extended to green 'tiger stripes' on a number of JG 27 aircraft. This one awaits its next alert scramble – or indeed the end of the game of 'skat' being played by the pith-helmeted pilots

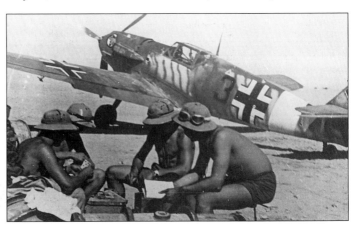

German pilots outright and his machine bellied in nearby, thus becoming the first virtually intact Bf 109F to be examined by the Allies. Later that day the Bf 109s bounced what had become a familiar target – a single tactical reconnaissance (TAC-R) Hurricane. These aircraft represented a great danger to the Afrika Korps on the ground, and they were regularly intercepted. A

heavy escort therefore had to be provided, and on this occasion it comprised both other Hurricanes and a top cover of Tomahawks. Oblt Rödel and Lt Schacht claimed one Curtiss each, although only a single Tomahawk of No 112 Sqn was actually lost.

Those German pilots who kept diaries noted their impressions of the desert war, and the kind of air combat they found. One wrote at this time, 'In nearly every case the initiative for attack lay with the German fighters. The rule was for two, four or six Bf 109s to meet anywhere between 12 to 20 (enemy) aircraft.'

Allied fighter controllers realised the vulnerability of the fighters, but the desert war invariably demanded low-level flights, often over very short short ranges. Much time would be lost, and precious fuel wasted, in climbing to high altitude if pilots had almost immediately to dive on their targets, which were usually troop convoys or panzers.

To counter Rommel's renewed bid to secure Tobruk the 8th Army struck first, and on 18 November Operation *Crusader* began. This, the largest Allied offensive yet seen in the desert, caught Rommel ill-prepared to repulse such an attack, and he was forced to yield ground.

ROMMEL'S CRITICISM

The *Jagdflieger* flew both *Frie Jagd* and escort sorties for *Stukageschwaderen* to thwart any gains the Allies made, but Rommel believed the Luftwaffe had not provided close enough support – indeed, Otto Heymer, the Koluft, was almost constantly in dispute with the Afrika Korps chief on this topic. *Crusader* brought the rift between the army and the Luftwaffe to something of a head, Rommel later having

Heymer court-martialled. Such actions did little to patch up the bad feeling which, not surprisingly, filtered down through the ranks.

Both Rommel's and Heymer's problems stemmed from the finite number of aircraft and tanks in-theatre. For their part, the *Jagdflieger* was obliged to whittle down the opposition in 'penny packet' hit and run attacks because there were never enough Bf 109s to offset the continual feed of Allied supplies into the region.

Movement of almost anything over the desert floor was invariably accompanied by a minor sandstorm, and a *schwarme* scramble, comprising a quartet of fighters with high revving engines all racing to get airborne in the shortest possible time, could hardly be hidden. Photographed in the early months of I./JG 27's deployment to Libya, these four Bf 109E-7/Trops are just seconds away from lift off here, each fighter dragging a characteristic sandy 'wake' behind it as it races across the Ain El Gazala strip at 110 kts

Among the I./JG 27 pilots who found the conditions of the desert air war much to their liking was Hans-Joachim Marseille. A crack shot with superb eyesight, Marseille nevertheless experienced a few mishaps, including getting this sizeable hole blown in his Bf 109E-4/Trop (one of a number of this model he was to fly – this machine still wears its old European scheme) in the spring of 1941. Marseille was then an Oberfähnrich (senior flight ensign), a lowly rank that, due to his high spirits, he seemed stuck with for many months *(Weal)*

Irksome and difficult, Stuka escort was the lot of JG 27 for a great deal of the time it spent in the desert. In this April 1941 view, Oblt Ludwig Franzisket, *Gruppe* adjutant, maintains station in his Bf 109E-7/Trop with one of 4./StG 2's Ju 87B-2s. This Messerschmitt was flown over from Sicily by Franzisket's when I./JG 27 deployed to Libya in April 1941, and it was decorated with 14 white kill bars on its fin at the time this photograph was taken later that same month

Hptm Erich Gerlitz, 2./JG 27's *Staffelkapitan*, flew 'Red 1' during May 1941, and scored 15 kills. His Bf 109E-7/Trop is seen here responding to an alert at Ain El Gazala

That the German fighters were able to inflict serious casualties on the enemy was shown on 20 November when I./JG 27 fell on an unescorted formation of nine No 21 Sqn SAAF Marylands and shot down four of their number in quick order – three of these were credited to Lt Hans-Arnold Stahlschmidt. This combat became known as 'Black Thursday', and was described in the following terms by the squadron diarist.

'The South Africans hardly knew what hit them. They had the impression of being "raked from front to stern by a whithering hail of cannon and machine gun fire". The Germans, a gunner later noted, were led by a master pilot who carried out his tactics "with faultless precision".

'The Marylands jettisoned their bombs, closed formation and dived to gain speed, but to no avail. Lt Stahlschmidt came in at them like a fury, attacking the rear sub-flight. The middle Maryland bore the brunt. A wing caught fire, but before it went down its gunners shot down a '109, then it ploughed in, killing all on board. The Germans then singled out the next flight. An American gunner attached to the squadron picked off a Messerschmitt trying to copy the tactics of Stahlschmidt, but his Maryland went down to the guns of another. Next attacked was the leading flight. One went down but the formation held together and the gunners fired back, took the attacker on and despatched him in flames.

'The Marylands were now fleeing at zero feet, and Lt Stahlschmidt bored in for his third kill, bringing down Maj Stewart's aircraft. Relentlessly, the Germans held on.

With four Marylands down they fastened onto the aircraft flown by Lt MacDonald. The "yellow (sic) nosed-demon" chased him for 60 miles and poured in "a hellish, raking fire" which wounded the bomber's gunner as the pilot "jinked, twisted, turned and dodged", before finally putting his aircraft down – a complete write-off with a hundred hits in it and both tyres shot to ribbons.' The diarist noted, with some despondency, that the unit had lost ten crews – forty men – since starting operations.

Early November recorded intense air combat before heavy rains hampered German units based on the forward airfields. After some three days waiting for drier conditions, the Bf 109s again clashed with Tomahawks. On the 20th honours were about even, two Bf 109s and four Ju 87s being lost for two Hurricanes and two Curtiss fighters. However, two days later JG 27 pilots hacked down ten Tomahawks and four Blenheims in return for six Bf 109s. The Allied fighters lost on 22 November had attempted to protect themselves by quickly performing the infamous defensive circle manoeuvre. The danger with this tactic was that if a Bf 109 pilot could penetrate the circle and make tighter turns, there was every chance of him picking off many targets. This was a risky business, however, for the enemy could theoretically range multiple guns against the Bf 109's four, and only pilots with the skill of Marseille tried it regularly. And being the most talented of all *experten*, it usually paid off handsomely for him!

Other American aircraft were now appearing in the desert to replace or supplement obsolescent types such as the Blenheim. One was the Douglas Boston light bomber which, although better all round than the old Bristol 'twin', was far from invulnerable. Just to prove this point to the RAF, Bf 109s caught a box of six unescorted Bostons heading westward into Libya on a bombing raid on 10 December and definitely shot down two, although three others also apparently failed to return to base. Five days previously the pendulum had briefly swung against I./JG 27 when *Gruppenkommanduer* Lippert was shot down and mortally wounded – he succumbed to his injuries a few days later in an Allied field hospital.

British forces had made good progress up to this point in *Crusader*, and by early December were pressing towards Tobruk. On the 7th, with the Axis forces retreating back past Ain El Gazala, JG 27 fled the airfield after some eight months' occupation, the unit flying to Martuba, with the recently arrived JG 53 fleeing back to Derna.

It was in December that the South African fighter units tried out tactics similar to those of the

Groundcrews run out to help a JG 27 pilot unstrap after another sortie over the frontline in September 1941. This E-7/Trop is rather unusual because it was one of only a handful of Bf 109Es to be resprayed in the solid tan scheme made popular with the later BF 109F/Trops. Despite their experience, the German pilots invariably found themselves up against superior numbers of Allied aircraft, a situation that steadily worsened for them as the months went by

To the victor, the spoils. Marseille took a professional hunter's interest in his victories, and often drove out to inspect the wrecks. This Hurricane Mk IIC of No 213 Sqn was downed by him in February 1942. Souvenir hunters have already stripped away the fabric which bore the roundel and code painted on. The rapidly ageing Hawker fighter was no match in a turning fight with a well-flown Bf 109E-7/Trop in the summer of 1941, let alone an F-4/Z Trop with an *Experten* at the controls some eight months later

Germans – No 1 Sqn flew a 'rhubarb' hit-and-run attack with pairs of fighters, finding, as had their adversaries, that the mutual cover offered was ideal for use against stragglers or larger formations. Hectic action, typical of this period, took place on 13 December and involved a number of Allied and German units; Marseille scored two and Obfw Herbert Krenzke and

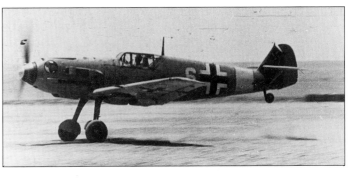

Oblt Gerhard Homuth one each, whilst a JG 53 pilot claimed a Beaufort, but the 'Pik As' *Geschwader* in return lost Lt Volkelmann. Also shot down on this day was Obfw Albert Espenlaub, a I./JG 27 *experten* with 14 kills who was now made a prisoner – he subsequently became one of the few German pilots to be shot by the Allies whilst attempting to escape.

It was now Derna's turn to be overrun by the advancing enemy, so JG 53 packed up and left. Before they went, however, the groundcrews scrawled a message of seasonal good cheer on the wall of the control room which read, 'We come back! Happy Christmas!' But fuel was now in critically short supply, hindering the *Jagdwaffe's* ability to perform offensive sweeps in support of the hard-pressed ground troops. They anxiously awaited the arrival of the Ju 52s. . .

MORE MALTA STRIKES

Albert Kesselring was, as *Oberfehlshaber Sud,* responsible for all Luftwaffe and flak units in the Mediterranean, and he welcomed additional strength to his command in November in the shape of Bruno Lortzer's *II.Fliegerkorps,* which had transferred in from central Russia. Besides the seasoned bomber *Gruppen* within this force, Lortzer brought with him all three *Gruppen* of JG 53, plus II./ JG 3. They would all be heavily involved in the new phase of bombing raids against Malta.

Based in Sicily for operations that commenced in mid-December, JG 53 was led by Maj Günther von Maltzahn, and among the unit's notable pilots were I. *Gruppenkommandeur* Hptm Herbert Kaminski, Oblt Freidrich-Karl 'Tutti' Müller (I./*Staffelkapitan*), Fw Herbert Rollwage

Scrambling into action in 'White 6', a pilot from 3./JG 27 demonstrates the perfect angle for the Bf 109E-7/Trop to achieve on the cusp of take-off. It might look easy, but any mistake at this point could be at least embarrassing, and at worst, fatal. A close examination of this shot reveals that the sliding doors that closed over the rudimentary air intake sand filter have been firmly shut prior to the launch. The pilot would then open the doors soon after take-off, thus allowing air to flow into the engine's supercharger unhindered

Reconnaissance was a vital element of the Luftwaffe's desert war and I./JG 27 was issued with examples of the Bf 109E-4/N variant, which had an oblique camera in a ventral housing. This excellent view of 'White 12', taken in June 1941 shows this to advantage, along with a drop tank to boost the basic 412-mile range. The 'N' element of the aircraft's designation was derived from its new high-compression Daimler-Benz DB 601N engine, which was capable of producing an extra 150 hp in emergencies

Despite its responsibility as the sole single-engined *Jagdwaffe* fighter *Geschwader* based permanently in North Africa, JG 27 only received the Bf 109F after units in Western Europe had been re-equipped. Like the *Staffel*, the Stab of II./*JG 27* also continued to fly the Emil well into the summer of 1941. When this shot was taken at Strumatal, in Bulgaria, in the early spring, Friedrichs were still months away for the unit. The Bf 109E-7 in the foreground is Wr-Nr. 4148, which belonged to *Gruppenkommandeur* Hptm Wolfgang Lippert. Parked alongside his aircraft is the suitably marked E-7 of the *Geschwaderkommodore*, Maj Wolfgang Schellmann *(Crow)*

A closer view of Lippert's machine, showing the impressive row of 21 markers for his victories up to the spring of 1941. He was to gain another four before being mortally wounded over Libya and dying as a PoW in late November that same year *(Crow)*

and Hptm Wolfe-Dietrich Wilcke. All these men had a substantial number of victories to their credit – indeed, most JG 53 pilots had scored kills at some point.

Malta operations had hardly begun before III./JG 53 was ordered from Sicily to Libya on 11 December to operate alongside a hard-pressed JG 27. Additional fighter strength in the shape of III./JG 27 was also despatched to North Africa, the new arrivals reuniting all three *Gruppen* of JG 27 for the first time since the desert fighting had begun. The *Jagdgeschwader* Kommodore, Maj Bernhard Woldenga, and his *Geschwaderstab*, accompanied the personnel and aircraft of III.*Gruppe*. Of the latter unit's *experten*, Oblt Erbo Graf von Kageneck was top scorer with 65 victories. JG 27 was more than ready to absorb fresh faces, for the pace of desert fighting had significantly quickened, and Allied air strength was beginning to present an increasingly difficult challenge. As earlier mentioned, combat of 13 December removed Obfw Albert Espenlaub from the strength of I.*Gruppe*, the pilot having claimed 14 victories between his arrival in April and his eventual death.

When III./JG 27 began operations, III./JG 53 rotated back to Sicily on 17 December, and soon after arriving on the Italian island fate struck a cruel blow when von Kageneck was mortally wounded on Christmas Eve. On the credit side, combat over Malta with Hurricanes boosted individual *Jagdfliegers'* scores, and significantly reduced bomber losses.

With supplies being the key to either side maintaining an advance or having to reduce the scale of fighting to conserve fuel, the desert armies were soon affected by events elsewhere. Thus, the Luftwaffe's new Blitz on the embattled island worked through fairly rapidly to Rommel's advantage as a fleet of Ju 52s criss-crossed the Mediterranean unmolested by the otherwise occupied RAF fighters on Malta. With fresh fuel for his panzers, he was able to advance once more, and by mid-January 1942 the British were on the retreat back towards Ain El Gazala. The ultimate prize was Tobruk, and by the 28th it was in German hands. The Luftwaffe's airfields at Martuba, Tmimi and Derna were also recaptured, leading one to wonder in what state the Germans found their former bases. At the latter strip, for example, more than 100 aircraft had been hastily abandoned by the Luftwaffe just the month before as the 8th Army advanced across Libya – that same force was now in full retreat.

Overhead, JG 27's pilots found a plethora of targets, and invariably made good use of their opportunities. On 14 January, despite reports that the RAF had at last started to adopt the 'finger four' fighter formation, II.*Gruppe* bounced Hurricanes of No 94 Sqn, and quickly despatched four of their number in short order – amazingly, they all fell to the guns of the same pilot, Uffz Horst Reuter. Fortunately for the Allies, his promising career was stopped just as it began to blossom as he was shot down and captured in May.

Homuth and Marseille, both of I./JG 27, and Otto Schulz of II.*Gruppe*, were among those pilots increasing their scores steadily. Marseille, in fact, rose to the position of top scorer in the desert when he shot down four enemy aircraft on 8 February, thus raising his tally to 40.

Experten continued to amass scores at a faster rate than their colleagues, a phenomenon not unique to the desert. Such was the espirit de corps of the *Jagdflieger* that these individuals were feted by the beginners, who strove to emulate their success, and assist them in achieving kills whenever possible. The race for top honours saw Homuth and Schulz close on Marseille's heels, the former reaching his 40th on 9 February. Three days later Schulz had a highly successful engagement with No 94 Sqn, although it might so easily have gone the other way for the Kittyhawks were actually strafing Martuba as he took off! As mentioned earlier, JG 27 had already taken a heavy toll of this unit only the month before when it had flown Hurricanes, and its new Kittyhawks fared little better, at least on this occasion. Five went down under Schulz's fire, one of which was flown by Sqn Ldr E M 'Imshi' Mason, a leading Allied ace of the theatre.

On 14 February the confusion that often reigned when an air battle developed was highlighted by RAF claims of 20 Axis fighters downed – at least 32 aircraft were involved in the mélee, and losses were restricted to three Bf 109s shot down and two damaged. For their feats of arms, Marseille and Schulz joined the ranks of JG 27's *Ritterkreuztrager* (Knight's Cross holders), which included Homuth, Rödel, Franzisket and Redlich. Schulz left II.*Gruppe* later that month to return to Germany for officer training.

During March 1942 the first Spitfires had arrived on Malta, but despite the respect with which the *Jagdflieger* treated the Supermarine fighter, there was no immediate change in British fortunes. Axis Bf 109s still found good hunting over and around the bomb-cratered island, the weary defenders doggedly resisting all attempts to annihilate them. Not that the Germans had all the luck – Lt Hermann Neuhoff, who had scored 40 kills up to this point with III./JG 53, was shot down in error by his own wingman and captured on 10 April, whilst on the 14th II./JG 3 lost its Kommandeur, Hptm Karl-Heinz Krahl, to Luqa airfield's anti-aircraft guns, which had by now become deadly accurate through constant practice!

By the autumn of 1941 II./JG 27 had received its first Bf 109F-2/Z Trops, and was on its way to North Africa to join I.*Gruppe*. This Friedrich has the unit's famous 'Berlin bear' emblem painted on its elegant nose contours, plus an unidentifiable personal motif beneath the cockpit. It was photographed already camouflaged, along with other Africa-bound F-4/Z Trops at Doberitz, in Germany, on 16 September – just a week prior to flying south *(Crow)*

A youthful-looking Lt Ernst Borngen of 5./JG 27 prepares for the flight to Libya. He is seen here fastening up his kapok-filled life jacket, which would have been a vital piece of kit should he have had to ditch en route. The *Gruppe* left Doberitz on 24 September *(Crow)*

At the same time as II.*Gruppe* were upgrading from Emils to Friedrichs, so too were individual *Staffeln* of I.*Gruppe* returning to Germany one at a time and trading up to the Bf 109F-4/Z Trop. Although there had been few complaints from pilots about the Bf 109E's ability to more than hold its own in North Africa, it was clear that the arrival of better Allied fighters and bombers for the Desert Air Force demanded the best possible counter from the *Jagdwaffe*. This I./JG 27 F-4/Z Trop was photographed moments after its DB 601E engine had coughed into life, the aircraft's mechanic retreating in its dusty slipstream clutching the powerplant's starter crank handle, which has clearly done its job

BRIEF LULL

With the ground fighting at another periodic stalemate, the German fighter units welcomed a reduction in the scale of combat. *Staffelen* alternated the responsibility of providing a readiness *rotte* or *kette* on different days, 'readiness' usually meaning take-off at 30 seconds' notice. Any reduction in vigilance could find the Germans caught on the ground by a bombing and strafing attack, or an incursion by the Special Air Service, who had begun to carry out raids behind enemy lines aimed at destroying both parked aircraft, and their all-important fuel dumps.

Not all early warning was localised though, a defence and interception centre having been established at Martuba which had a Gefechstand (Operations Room) and a Bodenstelle (Fighter Control) building. Like other links in the control network, this facility was built underground for maximum safety. Raids and interceptions were tracked on transparent paper laid over a large map which was used by the controller of the fighter R/T section, and he in turn passed 'orders of the day' to pilots already airborne. While the Luftwaffe and the Regia Aeronautica co-operated well in the air in Libya, the Italians nevertheless maintained a separate operations room and control system, with a German liaison officer attached.

In a theatre where, as already demonstrated around Crete, the German *Jabos* had achieved considerable success, JG 27 maintained a separate quick-reaction alert *Staffel* for this exacting task. Strikes by even a small number of fighter-bombers could put the enemy off-balance, and destroy or damage many aircraft in this war of attrition.

With the supporting nightfighter and *Zerstörer Gruppen*, which also had marked success in the early campaigns, *Fliegerführer Afrika* could, by May 1942, count on 92 Bf 109Fs of the three *Gruppen* of JG 27, plus eight *Jabos* based on Martuba's satellite fields. On 20 May III./JG 53 returned to Libya, and personnel changes due to postings saw Gustav

While the lightly-armed Bf 109F was not the complete answer to the *Jadgwaffe*'s needs, it was one of the most responsive variants of the entire series. In the desert there were few finer exponents of the Friedrich than Hans-Joachim Marseille, whose half-dozen or so personal aircraft marked as 'Yellow 14' became rightly famous. The Berlin-born *experten* they dubbed the 'Star of Afrika' is seen here climbing out of his first F-4/Z Trop, Wk-Nr. 12 593, in February 1942 after downing his 49th and 50th kills

Rödel taking command of II./JG 27. The despatch of *Gruppen* to Russia now left II./JG 53 as the only Bf 109 unit in Sicily, the shortfall in German fighter strength being made good with Italian units.

Renewal of offensive operations over Malta brought I./JG 77 back from Russia on 6 July, Hptm Heinz 'Prinzl' Bär having recently taken command of the *Gruppe*. With 120 victories to his credit, and a long list of medals and awards to match, Bär was the most highly decorated *experten* yet to grace the Mediterranean war front.

In May Rommel had moved against the British defences at Ain El Gazala and the French force defending Bir Hacheim. The battle, which involved most of the available ground forces on each side, continued into June, and by the 9th the Luftwaffe had flown 1030 sorties in support of the Afrika Korps, most of them to Bir Hacheim. Another 'maximum' effort was made by the Luftwaffe on the 10th, 124 Ju 87s and 76 Ju 88s attacking the fort complex, covered by 168 Bf 109s. The first Spitfires operating over the Western Desert were noted in the *Jagdfliegers'* combat reports on 10 June, their presence being seen as an ominous reminder of the escalation in Allied strength in the region. Nevertheless, the Bir Hacheim garrison surrendered on 11 June.

With his score on 91 by 15 June, Marseille was promoted to Kapitan of 3 *Staffel*, with 'Edu' Neumann becoming *Geschwaderkommodore* and Homuth taking over I. *Gruppe* at the same time. In company with III./JG 53, I./JG 27 now pursued the Allies, who were once again retreating. A string of contradictory movement orders then dogged JG 53, the end result being that they had aircraft spread all over the Mediterranean before, finally, the unit returned to Libya as an entity on 20 June.

By then, Marseille's score had risen to 101, and he was ordered to Berlin the receive the Swords for his *Ritterkreuz* from the *Führer* personally. As the first pilot to have scored 100 kills exclusively against the RAF, Marseille more than deserved this accolade, which had come only 12 days after the award of the Oak Leaves. He left Libya and took two months' leave, Marseille's departure being soured on the eve of his homecoming by the news that his rival Otto Schulz had been lost in combat with Kittyhawks the same day that the 'Star of Afrika' had scored his 101st kill.

When he met Hitler, Marseille was apparently given the chance to air his views on aspects of the cam-

Bf 109F-4/Z Trop of II./JG 27 at Sanyet in September 1942. The clean lines of this variant are readily apparent, as is the neat attachment of the centreline 300-litre drop tank. Both this airframe and the factory-coded Friedrich behind it had only recently arrived in Libya from Germany when photographed

Interesting comparison with the previous shot is the 'closed' intake of the Bf 109F's air filter. The drastic oil loss evident around the spinner and nose probably indicates that the pilot of this II./JG 27 fighter was grateful to have made a three-point landing in his weary mount!

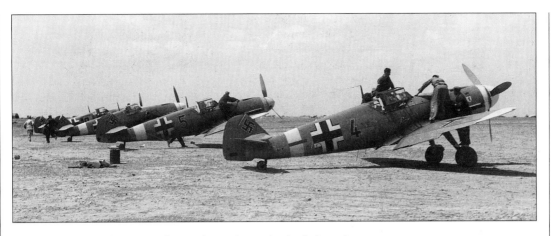

paign, and among the points discussed was the strained relations the Luftwaffe still had with Rommel. What effect these comments had on Hitler, or indeed on the prosecution of the war, is unknown – given the *Führer*'s pre-occupation with Russia it is assumed that the remarks were merely noted.

Events on the ground continued to move swiftly as Rommel forced the 8th Army back into Egypt, and his forces recaptured Tobruk on 19 June. By the 22nd I./JG 27 was based at Gambut and four days later at Sidi Barrani, where to their chagrin, the pilots found only one fuel bowser and no food. Sorties continued, however, and in combat Friedrich Körner shot down five and Stahlschmidt and Schroer three apiece. Vacating this less than hospitable base, the *Gruppen* moved to Bir El Astas on 27 June, before finally arriving at Fuka on the 29th. Again there was no fuel for the Bf 109s, and to cap it all a sandstorm broke the following day to ground the fighters, causing them to miss operations over El Alamein for a period.

Fuka was to remain JG 27's base for a time, as the Luftwaffe and Afrika Korps (as well as the British) strove to consolidate their respective forces. However, by this stage in the campaign the German fighters were all too easily located and bombed not only by the RAF, but also by the gradually burgeoning force of USAAF heavy bombers in-theatre. Despite the fact that the improved Bf 109G-2/Trop was also now being used by the *Jagdwaffe* in North Africa for the first time, many of these new aircraft were being flown by equally fresh pilots who had arrived in the Western

Part of a sequence of photographs taken of a *schwarme* of 5./JG 27 Bf 109Fs ('Black 4, 5, 3 and 10' from right to left) lined up for an *Alarmstart* take-off in early 1942 at Tmimi. This was possibly a special press/radio facility for war correspondents flown out from the 'Fatherland', although the 'alarm' itself was invariably real enough. In this shot the handful of 'black men' are inserting the hand cranks into the engines and slowly coaxing the DB 601s into life. Whilst this is occuring other groundcrew are preparing the pilots' seat straps and parachute harnesses within the cockpit

All hell breaks loose as engines rev up and air-beds and other loose objects left around by the pilots and groundcrew take flight. The *schwarme* is only seconds away from launching now, although the pilot of 'Black 4', seen here still sprinting to his machine, is going to have to 'shake a leg' if he wants to keep up with his fellow *Jagdflieger*!

At the other end of the line-up featured on the previous page, II.*Gruppe* groundcrew still await the signal to crank up the engine of 'Black 10'. This F-4/Z Trop has had its *Staffel* number crudely applied over the factory radio call letters

The elation of success in combat was invariably tempered by the saddness felt when comrades were lost, the pain being even more acute when an *experten* went down. Obfw Albert Espenlaub of 1./JG 27 crash-landed his Bf 109F-2/Trop behind enemy lines after taking hits in the radiator near El Adam on 13 December 1941. None too keen about becoming a PoW, he quickly attempted to escape and was promptly shot dead. Souvenir hunters have already stripped the fin of Espenlaub's kill markings

Desert straight from training schools. Not all of them had the guile of their predecessors, with many exhibiting signs of 'nervousness and inexperience', as one Allied observer put it. 'They opened fire at too great a range, and even at short range their shooting was bad.'

Such a situation was bound to occur as even an *experten* needed time to 'get his eye in', and to get a feel for the theatre in general, and the tactics of the opposition in particular. In the early days pilots had had months to perfect their technique, but by mid-1942 such a luxury was no longer possible. Day in, day out, the *Jagdflieger* were obliged to fly multiple sorties, most of which made contact with the enemy.

This situation was in stark contrast to the position of Allied fighter pilots, each of whom flew a set tour of duty which ended with an option to extend or be rested. While many RAF and Commonwealth pilots opted for a second or third tour, others returned home for a spell on training or staff duties. This system resulted in Allied pilots rarely suffering the 'combat fatigue' problems that began to plague the Luftwaffe by the second half of 1942. The only positive side to the German system was that individuals built up vast experience until, in all too many cases, the odds were too great for them to overcome. The death in action of so many of the finest *Jagdflieger* thus denied German pilot cadets much invaluable knowledge of the conditions they were likely to face.

A further factor was promotion. The wartime Luftwaffe was notoriously lax in this respect and, compared to the RAF for example, would regularly have a *Geschwader* (an RAF wing) commanded by the equivalent of a Flight Lieutenant. And although a low rank in fact carried more weight in the Luftwaffe than in the RAF, the responsibility of maintain-

ing a fighting unit in the frontline was a considerable burden to combat flying, which most commanding officers continued to perform. In many cases in the Allied air forces, promotion was followed by an encouragement to stay on the ground and impart one's expertise onto others. While not appreciated by those who would have preferred to remain operational, this system did, nevertheless, result in a better survival rate amongst novice and seasoned pilots alike, and spread experience broadly.

Rommel's offensive had once again taken the heat off Malta, which had even enjoyed the luxury of sending Spitfires across the Mediterranean to attack airfields on Sicily. August, therefore, settled into another lull while the combatants prepared for the inevitable next round.

By December 1941 III./JG 27 had joined the rest of the *Geschwader* in Libya. This view shows one of the unit's Bf 109Fs revving up, sans spinner, although the pilot would not attempt flight without it. Bf 109 spinners could often prove difficult to align following their removal to allow mechanics to get at the propeller reduction gear, and many bore witness to the groundcrews' frustrated efforts in achieving the desired fit. Engine runs without a spinner to check that all was working correctly saved time, but in turn ran the risk of sand ingestion

MONTGOMERY ARRIVES

It was one of the ironies of war that caused JG 27 to be directly responsible for placing the redoubtable Bernard Montgomery squarely in Rommel's path. On 7 August Uffz Bernd Schneider of II./JG 27 intercepted and destroyed the Bristol Bombay transport carrying Gen Gott, the designated new commander of the 8th Army, as he was being flown out to Egypt. Gott was killed in the ensuing crash, and his place was taken by Montgomery, whose leadership of the 'Desert Rats' soon became all too familiar to the Afrika Korps. Air combat took place around El Alamein prior to the attack commencing on 31 August when Rommel opened a fresh armoured thrust against the Allied lines at Adem el Haifa. Just a week before, Marseille had returned as the youngest Hauptmann in the Luftwaffe. He was then 22.

Hans-Joachim found plenty of action upon his return, the opposition having increased markedly in his absence. Several fierce air battles took place as the Desert Air Force strove to beat back the German panzers, as well as creating havoc in rear areas by destroying soft-skinned vehicles and rail transport vital to any major army offenive. Allied aircraft continued to attack the Luftwaffe on the ground, and a large number of fighters and bombers were lost to strafing attacks. Also, the Axis failure to neutralise Malta as an Allied base began working through to the detriment of air and ground forces alike, as supply lines from Italy began to come under constant attack.

Back in the cockpit after his short break, Marseille threw himself into this latest bout of aerial combat with exhausting vigour. On 1 September he was to have his greatest day.

When the 8th Army took the satellite strip at Daba in early 1942, they found a veritable graveyard of Luftwaffe aircraft, including this Bf 109F-4/Z Trop of 4./JG 27. Having to abandon dozens of aircraft in various states of repair as the Allies broke through the Afrika Korps defences greatly added to the *Jagdwaffe*'s mounting problems in the desert – these vital fighters were simply irreplaceable

END OF AN ERA

On 1 September II./JG 27 flew an early morning *Frie Jagd*. Take-off was at 0736 hours, and amongst the pilots of the six machines were Rödel, Sinner and Ofw Herbert Krenzke, the latter having 11 victories to his credit at that time. Two bomber formations comprising Baltimores and Bostons, covered by some 80 fighters, were intercepted. In attempting to get through to the bombers the *Jagdflieger* were instead obliged to engage the escorts. Sinner claimed two fighters shot down and Rödel one, although Krenzke's aircraft went down in flames, probably the victim of an attack by Spitfires of No 601 Sqn. The German victories were listed as 'Curtiss' fighters, their aircraft recognition on this, and numerous other occasions throughout the war, not being all that it should have been!

At 0840 four more II. *Gruppe* Bf 109s made contact with eight bombers and 20 fighters, and pilots sighted a second, similar size, formation at about the same time. No results were obtained from combat with this group, or with the 30 Hurricanes that were spotted ten minutes later.

This latter formation consisted of a No 80 Sqn TAC-R Hurricane, and its heavy escort of 12 similar machines from No 1 Sqn SAAF. They were attacked on their return flight by four Bf 109s at the German's chosen altitude of 6000 ft, whilst high above the dogfight at 15,000 ft, providing an extra layer of cover for the Hurricanes, were Spitfires of No 92 Sqn. The initial attack was followed up by the appearance of ten more Bf 109s, and the Hurricane pilots now had a hard fight on their hands, which was made much worse when their top cover was called away to attack a formation of Stukas. The No 1 Sqn pilots also heard this call but were too busy to respond.

The Ju 87 formation had an escort of aircraft from I./JG 27, and although the latter could not prevent the Spitfires engaging the dive bombers, those Stukas fired upon were only damaged, as were a number of Bf 109s.

Marseille, with Oblt Schlang acting as his wingman, flew escort to another Ju 87 formation between 0845 and 0900, in company with 13 other aircraft of I. *Gruppe* and 15 from III./JG 27. Attacked by 16 fighters, Marseille and Schlang

A considerable number of Bf 109s met their end in crash landings, such as this F-2 from II./JG 27. The 'Yellow 3' marking denotes that the abandoned fighter was possibly from to 6.*Staffel*. Aside from the obvious damage to the aircraft's propeller, the only other visible sign of it having been in combat is the small flak hole which the RAF officer is taking great pleasure in pointing out to the photographer *(Weal)*

Although the Bf 109 was hardly a type new to Allied technical evaluation teams, a number of virtually intact examples were retrieved for examination. This battered F-2 is being hitched to a light truck, probably for towing. Judging by the number of missing panels, the army beat the air force to this one! *(B Robertson)*

Although a number of fighter units spent brief periods in the desert, JG 53 'Pik-As' was one of the long-serving *Geschwader*, with all three of its *Gruppen* seeing duty in the area from late November 1941. In this view a Bf 109F is sharing a damp Sicilian airfield with two Fw 190As from II.JG 2, which were on their way to Tunisia from France. Note that the starter crank has been left in place ready for an *Alarmstart*

I.*Gruppe* JG 53 arrived in Sicily from Russia after II. and III.*Gruppen* in early October 1942. These Bf 109Fs are being checked prior to the next sortie over Malta. All have auxiliary fuel tanks, which became more widely used as the conflict worsened

turned into the enemy bounce, 'Jochen' opening fire in a left-hand turn. He immediately shot down the 'tail-end Charlie', plus a second machine, before returning to the Stukas and destroying a third fighter. Six Spitfires then latched onto him.

Taking his chance as the leader overshot, Marseille wrenched his Bf 109 into a port turn, shooting as he went. The Spitfire quickly fell away, the *experten* having used a mere 80 cannon shells and 240 rounds of machine gun fire to down four aircraft. Not all *Jagdflieger* were enjoying similar success, however, with both Fw Berben of I.*Gruppe* and Fw Gahr of III.*Gruppe* being lost, the latter becoming a PoW.

I.*Gruppe* again flew another 12 sorties at 1120 hrs as an escort force for Stukas, meeting two Allied bomber formations en route to the rendezvous. Marseille went for the fighter escort and fired on the leading formation of Kittyhawks, which promptly formed a defensive circle. Diving into the ring, he claimed two shot down as he broke up this manoeuvre. Latching onto the last fighter to break out of the ring, he clinically shot it down, then performed the same attack on a fourth and then a fifth Kittyhawk, the latter blowing up before it hit the ground. Marseille then nailed number six, his ninth for the day, in a steep left hand turn.

Climbing back on course for Fuka, his *schwarme* then saw another enemy formation above, and the ensuing attack brought Marseille his tenth kill, this last aircraft also blowing up.

His wingman, Schlang, fired on another wildly manoeuvring Kittyhawk during this same dogfight but missed, whereupon Marseille's left-hand turn took him

Few finer accolades were ever bestowed upon a German fighter pilot than those uttered by Generalmajor Adolf Galland during the summation of the career of Hans-Joachim Marseille. When the *Inspektor der Jagdflieger* called the top *experte* of the Western Desert campaign the 'unrivalled virtuoso of the fighter pilots', few of his contemporaraies would have disagreed.

It was fairly common for exceptional Luftwaffe pilots to make a slow, almost despairing, start to their combat careers, and Marseille was no exception. He joined I.(*Jagd*) LG 2 on 10 August 1940 and found dogfighting the RAF much to his taste, although his first combat with Hurricanes brought a realisation that he was not the only one with fine flying skills, good reflexes and a thirst for combat. This engagement saw Marseille in his Bf 109E-7 and an opposing pilot in a Hurricane Mk I fail to find the other's weak spot for a full four minutes, before the German gained a height advantage. He immediately dived on the RAF fighter and fired a short burst into him, which had the desired effect.

Marseille was to score a furhter five victories on the Channel front before the offensive against England was called off, surviving four ditching in the process. In late August he was transferred to 4./JG 52, and awarded the Iron Cross (1st Class) a month later. His later prowess over the desert began to emerge at around this time, although his CO, 'Macki' Steinhoff, took a less than benevolent view of Marseille's high spirits, which were often misconstrued as insubordination – this was one of the main reasons why he stayed a lowly Oberfahrnich for so long.

It was perhaps fortunate that Marseille was transferred to I./JG 27 at the end of 1940 as his new commanding officer, 'Edu' Neumann, soon recognised the qualities the man had in the only place it really mattered to the *Jagdwaffe* – at the controls of a fighter.

Thus it was that when JG 27

Marseille, wearing a favourite leather flying jacket, acknowledges the photographer as his most recent victory is recorded on the rudder of his first Bf 109F. The rows of kills were easily applied with the aid of a stencil held by the groundcrewman. A windsock hangs limply in the background

The 'Star of Afrika' is helped with his straps prior to another sortie commencing. This aircraft is believed to be the doomed Bf 109G-2 that suffered an engine fire which reulted in Marseille's death

moved to Libya, and its pilots found this new theatre much freer from official scrutiny, Marseille began to find his form.

While his opposite numbers in the RAF and SAAF gave due acknowledgement to the skill and tactics of the young *Jagdflieger*, it is doubtful if many of them knew the names of the pilots they were up against – Marseille became well-known at home in Germany, however. He also quickly took to the new Bf 109F, the *experten*'s mount. A superb exponent of deflection shooting, Marseille's economy in ammunition became little short of legendary. So well could he judge range, speed and angle, he was able to shoot down fighters with remarkably few cannon and machine gun rounds.

Marseille's combat record duly brought him the *Ritterkreuz* on 22 February 1942 (for 50 victories); the *Eichenlaub* on 6 June 1942 (75); the *Schwerten* on 18 June 1942 (101); and the *Brillianten* on 4 September 1942 (126). He was the only recipient of the *Brillianten* in JG 27 (it was never actually physically presented to him), although other pilots received the *Schwerten* and *Eichenlaub*.

In common with many long-serving *experten*, Marseille used many different aircraft during the course of his career. The exact number must, perforce, remain speculative, but the total is unlikely to be less than a dozen. From available records, it appears that Marseille lost four Bf 109E-7s during Channel Front operations with I.(J)/LG 2, one Bf 109E-7/Trop en route to Libya (marked as 'Yellow 13', rather than '14', the identification he carried on all his subsequent aircraft) and a second of this sub-type after scoring his first victories in it – he was shot down on 23 April 1941 by a Hurricane Mk I of No 73 Sqn flown by a Sous Lt Denis, and forced-landed behind German lines. An attack on Daba on 24 July destroyed Marseille's first Bf 109F-4/Z Trop, thus bringing his total of Messerschmitts lost to at least seven.

within extremely close range (100 yds) and he duly shot this machine down also – that made 11. Lt Remmer claimed a kill from another formation which then appeared, whilst Marseille flew back to base to be greeted by Generalfeldmarschall Albert Kesselring, commander of *Luftflotte* 2. The pilot was heartily congratulated on his morning's work.

Marseille was airborne again later in the afternoon, I. *Gruppe* having been briefed to provide both close and indirect escort for a large formation of Ju 88s. Flying as one of an 11-strong I./JG 27 formation, Marseille was ordered to provide the close cover. Between 1847 and 1853 hrs he claimed another five victories – all Hurricanes of No 213 Sqn. Stahlschmidt also claimed two and Lt Karl von Lieres u Wilkau one, again all from the now decimated No 213 Sqn.

At the end of this hectic day of virtually non-stop aerial combat Marseille put in a claim for 16 fighters destroyed, although the official records were later adjusted to show that a further victim had in fact fallen to his guns. No other pilot in the Luftwaffe had ever achieved such a score against the Western Allies, and this record was to stand alone in this respect right through to the end of the war. Overall, it was only bettered once by a pilot in Russia, who claimed 18 kills in one day.

During this period of heavy activity involving dozens of sorties by both German and Allied aircraft, air combat on 1 September was particularly intense and spread over a vast area. A few Allied bombers were lost to flak, but the fighters were unable to claim any shot down. Consequently, Rommel lost many vehicles and tanks, and was forced onto the defensive.

By September 1942 JG 27 were regularly meeting Spitfire Mk VB/Cs in combat over the frontline, then being bombed upon returning to their airfields by Hurricane Mk IICs on almost a daily basis. Despite this constant pounding, the Allies could expect a hard fight if the

Pleasing flying view of a *rotte* of I./JG 53 Bf 109F-2/Z Trops which both bear the distinctive engine exhaust streaking common to all desert Messerschmitts. The 'Pik-As' *Geschwader* was one of the most consistent in regard to the application of its famous black Ace of Spades badge on the nose.

A photograph that is not what it might at first appear to be! Oblt Hans-Joachim Heineke was actually part of 9./JG 27, but having previously flown with JG 53, he decided to adorn his F-4 with a small rendition of the 'Pik-As'. Note the prominant windscreen armour fitted to this aircraft *(Crow)*

Aircraft of 8./JG 53 pictured in Sicily in early March 1943. The well-decorated Bf 109F-4/Z Trop 'Kanonenboot' in the foreground belonged to the *Staffelkapitan*, Oblt Franz Schiess, whose 38 victories are marked on the rudder *(Crow)*

More Bf 109F wrecks under Allied scrutiny at an unknown base, which had excellent bomb-blast protection for the fighters. In the foreground is the aircraft flown by the III./JG 27 adjutant, whilst that at left bears a similar chevron marking. A III.*Gruppe* horizontal 'squiggle' appears on 'Yellow 4' to the right *(D Becker)*

Jagdflieger found themselves in a good tactical position. An air battle in which the Germans 'fought like men inspired' occured on 3 September, and 'some hard dicing took place' for over 20 minutes. A Tac-R Hurricane had been sent to make a survey of the frontline, and as usual it was covered by 24 other Hawker fighters from Nos 1 and 24 SAAF Sqns. These were quickly intercepted by a formation of 15 Bf 109Fs and Macchi C.200s/C.202s.

Earlier that same day Marseille and Stahlschmidt were part of a small force that had intercepted a dawn raid consisting of RAF Bostons and Baltimores, plus USAAF B-25s – total light bomber sorties for that day alone reached 200. They were covered by Desert Air Force Kittyhawks, Hurricanes and Spitfires, plus American P-40Fs of the 57th Fighter Group (FG). The *Jagdflieger*, in response, flew no less than 139 sorties, *Frie Jagd* sweeps accounting for all but 13 of this total. On this occasion the bombers were well protected, despite attacks by eight Bf 109s.

No 145 Sqn's Spitfire Mk VB pilots engaged both Bf 109s and Macchis during the same dogfight, and claimed to have damaged three, whilst in return losing one of their own – a pair of Kittyhawks from another unit were also lost. A pilot from this latter squadron destroyed two Bf 109s, whilst Marseille and Stahlschmidt each claimed three Curtiss fighters.

The confusion of a sortie such as this was summed up by Stahlschmidt in a letter he wrote to his family soon after the mission.

'Today I have experienced my hardest combat. But at the same time it has been my most wonderful experience of comradeship in the air. We were eight Messerschmitts in the midst of an incredible whirling mass of enemy fighters. I flew my 109 for my life. I worked with every gram of energy and by the time we finished I was foaming at the mouth and utterly exhausted. Again and again we had enemy fighters on our tails. I was forced to dive three or four times, but I pulled

up again and rushed back into the turmoil. Once I seemed to have no escape; I had flown my 109 to its limits, but a Spitfire still sat behind me. At the last moment Marseille shot it down, 50 metres from my 109. I dived and pulled up. Seconds later I saw a Spitfire behind Marseille. I took aim at the enemy – I have never aimed so carefully – and he dived down burning. At the end of the combat only Marseille and I were left in the dogfight. Each of us had three kills. At home we climbed out of our planes and were thoroughly exhausted. Marseille had bullet holes in his 109 and I had 11 hits in mine. We embraced each other, but were unable to speak. It was an unforgettable event.'

Coming from a pilot of the calibre of Stahlschmidt, the above comments serve to illustrate just how much Allied combat tactics, not to mention aircraft, had improved. That the strain of meeting ever more aggressive enemy fighter pilots was telling on the desert *Jagdflieger* was demonstrated all too clearly four days later on 7 September when I. *Gruppe* suffered a terrible loss. Four Bf 109s took off on a *Frie Jagd* at 1450 and headed for the Alamein area. The Axis pilots soon spotted their quarry – a Tac R Hurricane. In deference to the anticipated reaction of the Bf 109s, the lone Allied fighter was covered by a strong escort of Hurricane Mk IICs from both Nos 33 and 213 Sqns. What the Germans overlooked on this occasion was that their own tactics were now being copied by the RAF – up sun were Spitfires Mk VCs of No 601 Sqn.

Boring in towards the Hurricanes, the *Jagdflieger* selected their targets and opened fire. This time things did not go at all as planned, for the British fighters turned into the attack and promptly claimed two Bf 109s. The downed aircraft were flown by Lts von Lieres u Wilkau (24 kills) and Stahlschmidt (59 kills). The former managed to survive a torrid crash-landing, but 'Fifi' Stahlschmidt was not so fortunate. The victor of many exhausting combats in over 400 operational flights was dead. Coming only the day after the death of Ofw Gunther

The Allied troops obviously found 'Yellow 4', noted in the previous photograph, of enough interest to place it on a flat-bed trailer for transportation away for further inspection. Note the fighter's white-walled tailwheel, the paint having been applied in an effort to deflect the sun's ray, and thus protect the tyre rubber. Once cleared of wrecks, the dispersal area would soon be occupied by Allied aircraft. A number of strips changed hands on more than one occasion *(D Becker)*

Placing an airfield 'under new management' could be accomplished very quickly, and numerous German bases were simply abandoned with little attempt to disable flyable aircraft. This Bf 109F-4 of I./JG 53 was one of many that fell into Allied hands in the steamrolling advance across Libya following the breakthrough at El Alamein. Various bits of canopy framing and windscreen are strewn around the rear of this virtually intact fighter *(D Becker)*

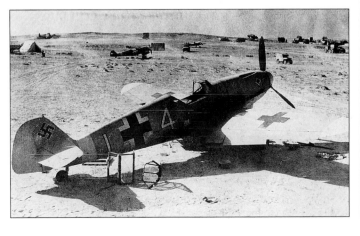

ORDER OF BATTLE

Despite the crippling losses of irreplaceable personnel, the Luftwaffe was at the peak of its operational aircraft strength on all fighting fronts in September 1942. On the 20th of that month the following fighter units were based in the Mediterranean:

Luftflotte 2:

Fliegerführer Afrika

Unit	A/C	Base	Strength	Serviceable
Stab./JG 27	Bf 109F	Sanyet/Quotaifiya	3	(2)
I./JG 27	Bf 109F	Turbiya	28	(15)
II./ JG 27	Bf 109F	Sanyet	26	(16)
III./JG 27	Bf 109F	Sanyet/Quasaba	28	(18)
III./ JG 53	Bf 109F	Quasaba East	27	(14)

Jabo Staffel Afrika

	Bf 109F	Quasaba East	27	(14)

Also III./ZG 1 at Quasaba East with 27 (14) Bf 109s and 8./ZG 26 at Barce/Derna with 12 (5) Bf 110s

II.Fliegerkorps – Sicily

Stab./ JG 53	Bf 109F	Comiso	5	(5)
II./JG 53	Bf 109F	Comiso	32	(25)
I./JG 77	Bf 109F	Comiso	34	(25)

X.Fliegerkorps – Greece and Crete

Jagdkommando JG 27	Bf 109F	Kastelli	15	(7)

Also III./ZG 26 (less 8.*Staffel*) had 37 (14) Bf 110s at Kastelli

Steinhausen, (another *experten* with 40 victories to his credit), this was a savage blow to the morale of I. *Gruppe*. Over a different sector, Marseille shot down two on this fateful day, followed by a further pair on the 11th, seven on the 15th and another seven on the 26th. His score for the first three-and-a-half weeks of September totalled 52 kills!

However, this frenetic pace was beginning to tell, even on such an outwardly calm character as Marseille. By now he had almost overtaken Stahlschmidt's astounding total of 59 kills in the space of just three weeks, but the continual loss of his comrades shook him, for there was now an aggressiveness in the tactics flown by the Allies that had not been seen before. Marseille's tally for September 1942 eventually rose to 61, following another seven-aircraft haul on the 28th. When he landed after the last combat of that day, he was visibly drained by the exertions of flying, and on the point of collapsing from exhaustion. Marseille

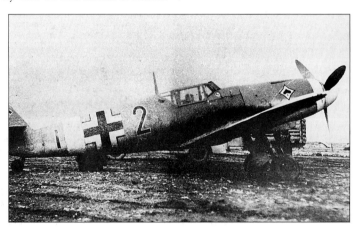

Whilst most captured aircraft were simply left to rot, the odd few were flown by the occupying forces, and some Allied fighter squadrons used Bf 109s as fast communication hacks. This ex-I./JG 53 example was taken over by No 2 Sqn SAAF at Martuba in late 1942 *(D Becker)*

The coding on this Bf 109F-4 denotes that it became the war booty of No 1 Sqn SAAF. Use of a question mark in place of an individual aircraft letter was common amongst frontline squadrons in North Africa, and quite appropriate in the circumstances for a captured machine *(D Becker)*

Shot down southwest of Zarzis on 20 February 1943, this Bf 109G-2/Trop 'Black 14' was being flown at that time by a Lt Wernicke of the little-known 2(H)/14 reconnaissance unit – the pilot escaped unhurt. The bulged ventral fairing for the camera installation can be made out below the *Balkenkreuz*

had just experienced a 15-minute tussle with a formation of Spitfires, during which he shot down his final victim, thus bringing his tally to 158 – far more than any other pilot achieved over the desert.

There was nothing spectacular about 30 September for I./JG 27 in terms of combat flying, I. and III. *Gruppen* having been briefed to escort Stukas yet again, but on this occasion the *Jagdflieger* made no contact with enemy fighters. Marseille was flying a brand new Bf 109G-2/Trop, and while returning to base he reported that his engine had caught alight following the fracturing of an oil line. Seconds later the other members of his flight noticed that the aircraft was beginning to smoke badly. Marseille was frantically urged to bale out, and rolling the Gustav onto its back, he fell away from the now-burning machine. As he did so his body struck the tailplane. Horrified, the other pilots watched their leader plunge earthward, his parachute canopy only partially streamed. Marseille hit the ground not far from his aircraft, sustaining fatal injuries upon impact with the desert floor.

So bad an effect did the death of Marseille, Stalhschmidt and Steinhausen have on JG 27 that I. *Gruppe* was immediately pulled out of North Africa and ordered to Sicily. Understandably, morale amongst both pilots and groundcrew sank to an all-time low. Adolf Galland visited Fuka shortly after these terrible events to confer with *Fliegerführer Afrika*

on the situation the *Jagdflieger* now faced. The news was not encouraging, with Galland being told that the Germans could muster only 80 fighters, as opposed to the Allies who now had more than 800.

The war, meanwhile, raged at even greater intensity. Fuka was hit by fighter-bombers on 9 October, Baltimores and Kittyhawks bombing and strafing at will. The latter adopted a line abreast formation for their attack, their fire scything across the southern dispersal area where the Bf 109 force was parked. Lying only five minutes flying time from Alamein, Fuka was vulnerable to such indignities, and this was not the first time it had been hit. However, line abreast strafing was a new tactic for the Allies, and it resulted in 13 of III./JG 53's 27 Bf 109F-4s being written off.

MALTA – AGAIN

In order to ease the pressure on Rommel, the Luftwaffe planned to deliver a series of blows against Malta that would, it was fervently hoped, remove this thorn in the side of the Afrika Korps once and for all. RAF anti-shipping strikes from the island were achieving much success, and thousands of tons of vital supplies had already been lost. On 11 October the first in a series of raids took off from Sicilian airfields.

Respite from the attentions of the *Stuka* and *Kampfgeschwaderen* had enabled Malta to profit from the general Allied build up in the region, and for defence against air attacks where there had once been a handful of antiquated Hurricane Mk Is there were now Spitfire Mk

Indicating a fair degree of cannibalisation, this anonymous I.*Gruppe* Bf 109F has had a wing from another machine fitted judging by the different style of *Balkenkreuz* marking. Empty machine gun belts and the remains of a wing cannon lay broiling on the desert floor alongside this long-since forgotten fighter *(D Howley)*

Unlike the sad hulk of the F-2 above, this II./JG 51 Bf 109G-1 was still very much alive when it was photographed at Catania, in Sicily, in November 1942, en route to the Libyan frontline. The pilot had just completed a post-maintenance check flight when this shot was taken *(Robertson)*

Yet another JG 53 Bf 109F-2/Z Trop left to its fate when the ground war came too close for comfort in late October 1942. This almost new machine was previously operated by III.*Gruppe (Robertson)*

Comiso, Sicily, October 1942. A crowded flightline shows Bf 109G-2s of 6./JG 53 (Wr-Nr. 10522 in the foreground), a Bf 110C of *Aufkl.Gr* (*H*) 41 and assorted Ju 52/3m transports. At the time II./JG 53 was preparing to depart for Tunisia. To avoid the fighter cockpits getting too hot the groundcrews have draped tarpaulins over each aircraft *(Weal)*

VB/Cs. In six raids involving Ju 88s, the fighter escorts were stepped up, although the final strike inexplicably had no cover at all. As a result, this mission suffered the worst casualties of all, losing four aircraft.

The third raid was escorted by I./JG 77 and Stab./JG 53. Two Spitfires were claimed by JG 77, one each by Uffz Schlick and Oblt Freytag, and a third by JG 53. Just as Marseille had been the 'Star of Afrika', so Siegfried Freytag became known as 'Die Stern von Malta' for his combat success. His tally rose to 73 over Malta, and he finihed the war with 102 kills.

The Ju 88s managed to drop their bombs on every raid, and although they were dispersed over a wide area, a number of RAF fighters were hit. For a week the Germans sent out sorties on successive days, and losses began to steadily mount despite the best efforts of I./JG 27, II./JG 53 and I./JG 77. It was clear to the crews involved that the defending forces were more than capable of warding off a far greater offensive. Over the seven-day Blitz, 34 Ju 88s and 12 Bf 109s were destroyed, plus a further 18 aircraft damaged. In return, the RAF admitted that 23 Spitfires were shot down and 20 crash-landed. In a similar situation to that which had existed over the Channel in 1940, only 12 RAF pilots were killed, however, whereas Axis losses invariably meant aircrew being captured, even if they came to earth uninjured. The Luftwaffe had left their assault too late.

THE COLOUR PLATES

This colour section profiles some of the most famous Bf 109s flown by the legendary *experten* in North Africa and the Mediterranean, plus lesser known examples that have never previously been illustrated. The artworks have all been specially commissioned, and profile artists Chris Davey, John Weal and Keith Fretwell, plus figure artist Mike Chappell, have gone to great pains to portray both the aircraft and the pilots as accurately as possible following much in-depth research.

1
Bf 109G-1 'White 11', flown by Oberleutnant Julius Meimberg, *Staffelkapitan* 11./JG 2, Bizerta/Tunisia, November 1942

2
Bf 109F-4/Z Trop 'White Chevron/Triangle', flown by Hauptmann Karl-Heinz Krahl, *Gruppenkommandeur* II./JG 3, San Pietro/Italy, April 1942

3
Bf 109F-4/Z Trop 'Yellow 3', flown by Unteroffizier Franz Schwaiger, 6./JG 3, Castel Benito/Libya, February 1942

4
Bf 109G-6/Trop 'Black Double Chevron', flown by Major Franz Beyer, *Gruppenkommandeur* IV./JG 3, San Severo/Italy,
August 1943

5
Bf 109E-7 'White 15', flown by Feldwebel Karl Laub, 7./JG 26, Ain El Gazala/Libya,
June 1941

6
Bf 109E-7 'White 12', flown by Oberleutnant Joachim Müncheberg, *Staffelkapitan* 7./JG 26, Gela/Sicily,
February 1941

7
Bf 109G-1 'Black 1', flown by Oberleutnant Hans-Jürgen Westphal, *Staffelkapitan* 11./JG 26, Trapani/Sicily,
November 1942

8
Bf 109F-4/Z Trop 'White Chevron A-bar', flown by Hauptmann Werner Schroer, Acting Adjutant JG 27, Martuba/Libya, circa November 1942

9
Bf 109F-4/Z Trop 'Black Chevron T', flown by Oberleutnant Rudolf Sinner, Technical Officer JG 27, Martuba/Libya, circa April 1942

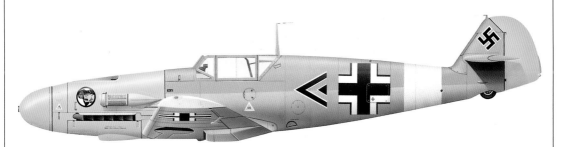

10
Bf 109F-2/Trop 'Black Chevron/Triangle', flown by Hauptmann Eduard Neumann, *Gruppenkommandeur* I./JG 27, Martuba/Libya, circa December 1941

11
Bf 109E-7/Trop 'Black Chevron A', flown by Oberleutnant Ludwig Franzisket, *Gruppen*-Adjutant I./JG 27, Castel Benito/Libya, April 1941

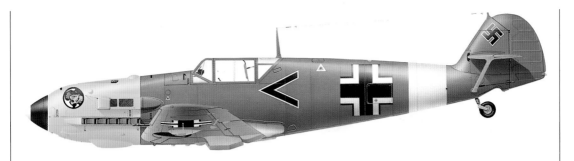

12
Bf 109E-7/Trop 'Black Chevron', flown by Oberleutnant Ludwig Franzisket, *Gruppen*-Adjutant I./JG 27,
Ain El Gazala/Libya, circa October 1941

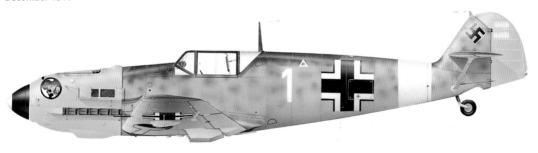

13
Bf 109F-2/Trop 'White 11', flown by Oberfeldwebel Albert Espenlaub, 1./JG 27, Martuba/Libya,
December 1941

14
Bf 109E-7/Trop 'White 1', flown by Oberleutnant Karl-Wolfgang Redlich, *Staffelkapitan* 1./JG 27, Ain El Gazala/Libya,
July 1941

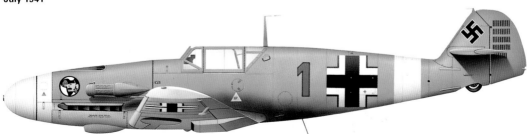

15
Bf 109F-4/Z Trop 'Red 1', flown by Leutnant Hans-Arnold Stahlschmidt, *Staffelkapitan* 2./JG 27, Quotaifiya/Egypt,
August 1942

16
Bf 109F-2/Trop 'Yellow 1', flown by Oberleutnant Gerhard Homuth, *Staffelkapitan* 3./JG 27, Martuba/Libya, February 1942

17
Bf 109F-4/Z Trop 'Yellow 14', flown by Leutnant Hans-Joachim Marseille, 3./JG 27, Martuba/Libya, February 1942

18
Bf 109F-4/Z Trop 'Yellow 14', flown by Leutnant Hans-Joachim Marseille,3./JG 27, Tmimi/Libya, May 1942

19
Bf 109F-4/Z Trop 'Yellow 14', flown by Oberleutnant Hans-Joachim Marseille, *Staffelkapitan* 3./JG 27, Ain El Gazala/Libya, June 1942

20
Bf 109F-4/Trop 'Yellow 14', flown by Hauptmann Hans-Joachim Marseille, *Staffelkapitan* 3./JG 27, Quotaifiya/Egypt,
September 1942

21
Bf 109F-4/Trop 'Black Double Chevron', flown by Hauptmann Wolfgang Lippert, *Gruppenkommandeur* II./JG 27,
Ain El Gazala/Libya, November 1941

22
Bf 109G-4/Trop 'White Triple Chevron 4', flown by Hauptmann Gustav Rödel, *Gruppenkommandeur* II./JG 27,
Trapani/Sicily, April 1943

23
Bf-109F-4/Z Trop 'Black Chevron', flown by Oberleutnant Ernst Düllberg, *Gruppen*-Adjutant II./JG 27, Tmimi//Libya,
May 1942

24
Bf 109F-4/Z Trop 'White 12', flown by Oberfeldwebel Franz Stiegler, 4./JG 27, Quotaifiya/Egypt,
August 1942

25
Bf 109F-4/Z Trop 'Yellow 2', flown by Oberfeldwebel Otto Schulz, 6./JG 27, Tmimi/Libya,
May 1942

26
Bf 109F-4/Z Trop 'Yellow 1', flown by Oberleutnant Rudolf Sinner, 6./JG 27, Tmimi/Libya,
June 1942

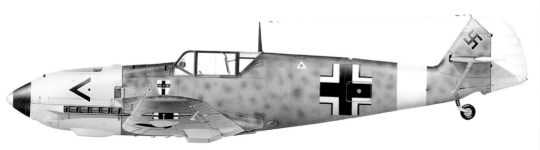

27
Bf 109E-4 'Black Chevron/Triangle', flown by Hauptmann Max Dobislav, *Gruppenkommandeur* III./JG 27, Sicily,
May 1941

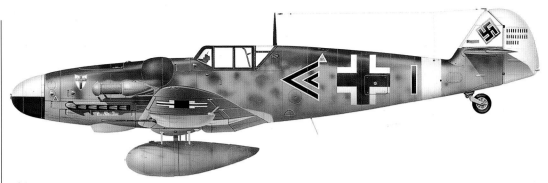

28
Bf 109G-6/R6 Trop 'Black Double Chevron', flown by Hauptmann Ernst Düllberg, *Gruppenkommandeur* **III./JG 27,**
Argos/Greece, circa October 1943

29
Bf 109G-6/R6 Trop 'White 9', flown by Oberleutnant Emil Clade, *Staffelkapitan* **7./JG 27, Kalamaki/Greece,**
January 1944

30
Bf 109E-7/Trop 'Black 8', flown by Leutnant Werner Schroer, 8./JG 27, Ain El Gazala/Libya,
April 1941

31
Bf 109G-2/Trop 'Red 1', flown by Hauptmann Werner Schroer, *Staffelkapitan* **8./JG 27, Rhodes,**
circa February 1943

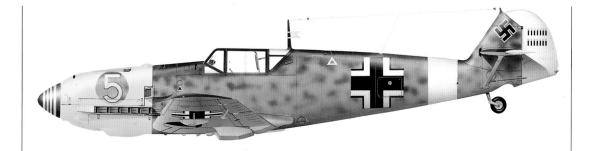

32
Bf 109E-4 'Yellow 5', flown by Oberleutnant Erbo Graf von Kageneck, *Staffelkapitan* 9./JG 27, Sicily,
May 1941

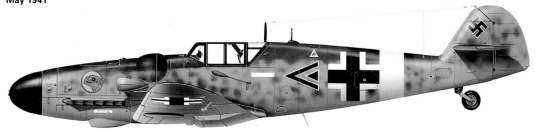

34
Bf 109G-6 'Black Chevron/Triangle', flown by Hauptmann Karl Rammelt, *Gruppenkommandeur* II./JG 51, Tuscania/italy,
circa February 1944

35
Bf 109G-2/Trop 'White 5', flown by Feldwebel Anton Hafner, 4./JG 51, Bizerta/Tunisia,
November 1942

36
Bf 109G-6/Trop 'White 12', flown by Oberfeldwebel Wilhelm Mink, 4./JG 51, Tuscania/Italy,
circa January 1944

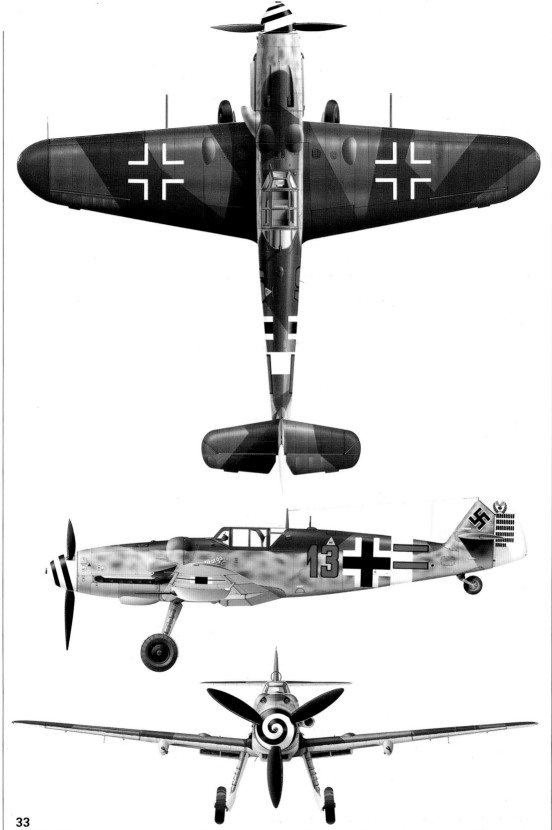

33
Bf 109G-6/R6 'Red 13', flown by Feldwebel Heinrich Bartels, 11.JG/27, Kalamaki/Greece, circa September 1943

37
Bf 109F-4 'Black Chvron/Triangle and Bars', flown by Oberstleutnant Günther Freiherr von Maltzahn, *Geschwaderkommodore* JG 53, Comiso/Sicily, circa May 1942

38
Bf 109G-6/R6 'Black Double Chevron', flown by Major Jürgen Harder, *Gruppenkommandeur* I./JG 53, Maniago/Northern Italy, March 1944

39
Bf 109G-2/R1 Trop 'Yellow 13', flown by Leutnant Wilhelm Crinius, 3./JG 53, Bizerta/Tunisia, January 1943

40
Bf 109G-4 'Yellow 7', flown by Oberleutnant Wolfgant Tonne, *Staffelkapitan* 3./JG 53, Bizerta/Tunisia,

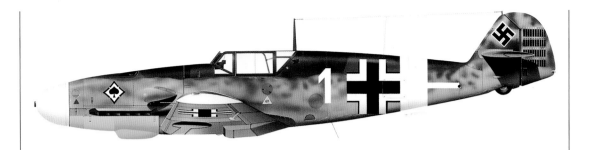

41
Bf 109F-4 'White 1', flown by Oberleutnant Gerhard Michalski, *Staffelkapitan* 4./JG 53, Pantelleria,
July 1942

42
Bf 109F-4/Z 'Black 1', flown by Hauptmann Kurt Brändle, *Staffelkapitan* 5./JG 53, Comiso/Sicily,
circa February 1942

43
Bf 109F-4 'Black 2', flown by Oberfeldwebel Herbert Rollwage, 5./JG 53, Pantelleria,
August 1942

44
Bf 109G-6 'Black 2', flown by Oberfeldwebel Herbert Rollwage, 5./JG 53, Trapani/Sicily,
July 1943

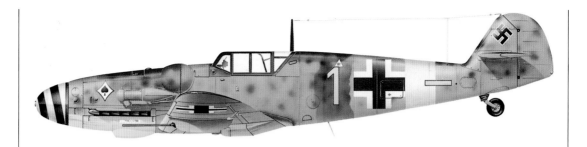

45
Bf 109G-6/R6 Trop 'Yellow 1', flown by Hauptmann Alfred Hammer, *Staffelkapitan* 6./JG 53, Cancello/Italy,
August 1943

46
Bf 109G-6 'Black Double Chevron', flown by Major Franz Götz, *Gruppenkommandeur* III./JG 53, Orvieto/Italy,
circa January 1944

47
Bf 109F-4/Z Trop 'White 5', flown by Leutnant Jürgen Harder, 7./JG53, Martuba/Libya,
June 1942

48
Bf 109G-4/Trop 'White 1', Hauptmann Jürgen Harder, *Staffelkapitan* 7./JG 53, Trapani/Sicily,
February 1943

49
Bf 109F-4 'White 2', flown by Leutnant Hermann Neuhoff, 7./JG 53, Comiso/Sicily,
March 1942

50
Bf 109G-4/Z Trop 'Black 1', Oberleutnant Franz Schiess, *Staffelkapitan* 8./JG 53,
Tunis-El Aouina/Tunisia, circa February 1943

51
Bf 109F-4/Z Trop 'Yellow 1', flown by Oberleutnant Franz Götz, *Staffelkapitan* 9./JG 53, Martuba/Libya,
circa June 1942

52
Bf 109F-4 'Black Double Chevron', flown by Leutnant Heinz-Edgar Bär, *Gruppenkommandeur* I./JG 77, southern Italy,
July 1942

53
Bf 109G-4/Trop 'Black Chevron', flown by Leutnant Heinz-Edgar Berres, *Gruppen*-Adjutant I./JG 77, Matmata/southern Tunisia, circa January 1943

54
Bf 109G-6 'White 1', flown by Oberleutnant Ernst-Wilhelm Reinert, *Staffelkapitan* 1./JG 77, southern Italy, circa August 1943

55
Bf 109F-2/Trop 'White 3', flown by Unteroffizier Horst Schlick, 1./JG 77, Bir El Abd/Egypt, November 1942

56
Bf 109E-4 'Black Chevron/Triangle', flown by Hauptmann Herbert Ihlefeld, *Gruppenkommandeur* I.(J)/LG 2, Molaoi/Greece, May 1941

57
Bf 109G-6/Trop 'Black 4', flown by Sottotenente Giuseppe Ruzzin 154a *Squadriglia*, 3° *Gruppo CT* (RA), Comiso/Sicily, July 1943

58
Bf 109G-6/Trop 'White 7', flown by Tenente Ugo Drago, Comandante 363a *Squadriglia*, 150° *Gruppo CT* (RA), Sciacca/Sicily, May 1943

59
Bf 109G-10/AS, flown by Maggiore Adriano Visconti, Comandante I° *Gruppo Caccia* (ANR), Lonate Pozzolo (Varese)/Italy, February 1945

60
Bf 109G-6 'Black 7', flown by Capitano Ugo Drago, Comandante 4a *Squadriglia*, II° *Gruppo Caccia* (ANR), Aviano/Italy, November 1944

1
Oberleutnant Hans-Joachim
Marseille in June 1942

2
Jagdflieger in tropical flying gear,
Libya, July 1941

3
Leutnant, circa early winter 1942 in
Libya

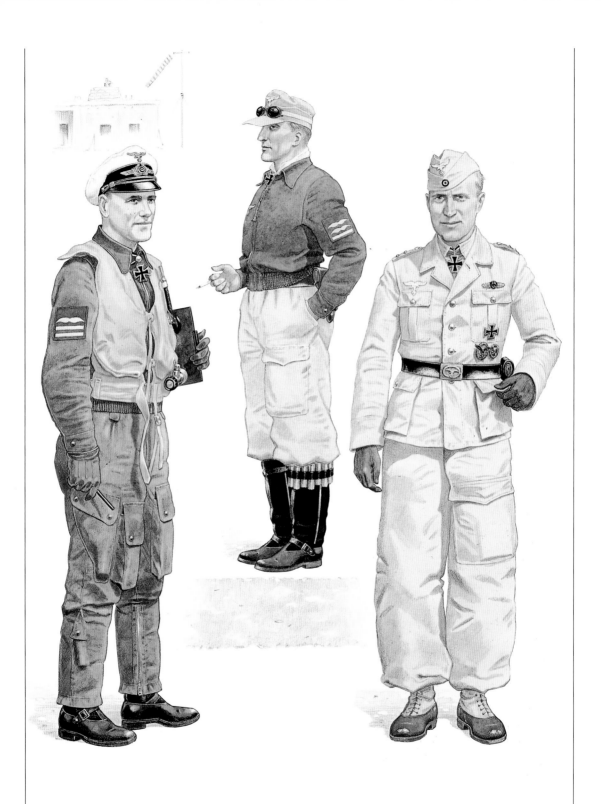

4
Major Joachim Müncheberg of 7./JG
26 on Sicily in spring 1941

5
Oberleutnant wearing a mix of Afrika
Korps and Luftwaffe clothing

6
Oberfeldwebel Otto Schultz of II./JG
27 in late 1941

NEW ADVERSARIES

With their supplies reaching North Africa in steadily increasing quantities thanks to much improved air and naval protection of convoys, the British forces rallied under Montgomery, who moved against Rommel only when the 8th Army was strong enough for a decisive offensive. When the battle of El Alamein began on 23 October, the Germans reeled under the combined onslaught of artillery fire, tanks and co-ordinated air support, and soon fell back. By the end of the month, the Afrika Korps was clearly on the defensive as German air support had been seriously weakened. Fighters from the Stab, II. and III./JG 27, III./JG 53 and the *Jabos* of I./Sch.G 2, nevertheless contributing to the land battle as much as they were able. I./JG 27 returned to the fray on the 27th flying in from Sicily with I./JG 77, whereupon the battle-weary III./JG 53 pulled back across the Mediterranean.

Plans now called for JG 27 to be entirely replaced by JG 77, and III.*Gruppe* duly arrived from the Eastern Front, to be followed by the *Geschwaderstab* under the command of the redoubtable Major Joachim Müncheberg. Since his earlier successes almost 18 months before, Müncheberg's score had risen to over 100, and his pilots were generally in excellent spirits after a sucessful period in Russia.

That the North Africa theatre was both unpredictable and now far more dangerous than the Eastern Front was quickly shown to III./JG 77 when on 29 October, just a day after their arrival in-theatre, *Staffelkapitan* Hptm Wolf-Dietrich Huy (a 40-victory *experte*) succumbed following a fierce dogfight with a formation of Spitfires – he was duly made a PoW. Also lost to the strength of I./JG 27 at the same time was another North African veteran, *Staffelkapitan* Hptm Ludwig Franzisket being shot down and suffering a badly broken leg in the ensuing crash. He had scored 37 kills in the desert up to this point.

November saw the Germans still retreating in the face of the Allied offensive, and so serious had the situation become that Rommel prepared

Pressurised Bf 109G-1 'Yellow 8' of 3./JG 53 was photographed in a makeshift revetment made of hay bails in Tunisia in early 1943. This machine was one of many ex-11./JG 2 aircraft which were absorbed by JG 53 following the former *Staffel*'s re-equipment with up-gunned Bf 109G-1/R6s in Germany. These specialized high-altitude interceptors were the first mass-produced pressurised Bf 109s, and they were sent to North Africa to deal with the USAAF's B-24 raids that were then just commencing. This aircraft was one of the first G-1s in-theatre, and as such boasts only the standard armament fit of two MG 131 machine guns in the fuselage and a solitary MG 151 20 mm cannon firing through the propeller hub – hardly enough to stop a well-armoured Liberator. The R6 version, however, saw a single podded MG 151 bolted beneath each of the G-1's wings, this formidable weapon being more than capable of stopping a multi-engined bomber in its tracks with only a handful of bursts *(Weal)*

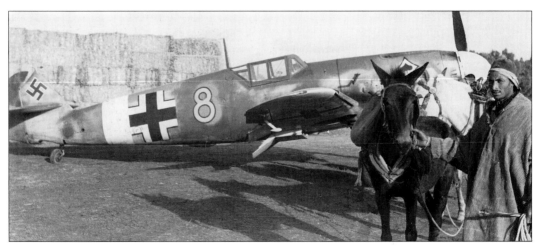

to abandon Egypt altogether and withdraw into Cyrenacia. The *Jagdwaffe* also began to move back, briefly occupying familiar airfields en route – Sidi Barrani, Menastir and Gambut. With each hasty departure westwards, more and more fighters that could not be evacuated due to unserviceablity were left behind. II./JG 27, reduced to only three serviceable Bf 109G-2s during the hurried retreat, was re-equipped with new Bf 109G-6s at Gambut, and was ready for operations again by 6 November.

A week or so later, the *Geschwader* was ordered out of North Africa, thus bringing to a close the most momentus period in its entire history. Not all personnel returned with the Stab and I. *Gruppe* to Germany, however, II. and III. *Gruppen* dispersing to Crete and Greece respectively.

JG 77, which had absorbed all of JG 27's remaining Bf 109s, followed the retreating armies. The Germans had, meanwhile, received news of the Allied *Torch* landings in French North Africa, which brought with it the potentially grim portent that eventually the Afrika Korps could be sandwiched between two advancing armies. By December further re-equipment of the Luftwaffe had taken place in the form of Hptm Anton Mader's II./JG 77 from Russia. This move enabled the almost decimated II./JG 27 to evacuate to Italy, which was one of the more popular decisisons made by the high command from the *Geschwader's* viewpoint after having endured weeks of retreating!

COUNTERING *TORCH*

Having scored their 1000th victory of the war soon after arriving in North Africa, pilots of I./JG 77 quickly got to grips with their opponents in this new war zone. II. *Gruppe* found that the Allies were more than capable of shooting down their aircraft, even with *experten* at the controls, and a number of Bf 109s were lost soon after the unit began operations.

The Axis forces now turned their attention to French Morocco and Algeria, and part of the Luftwaffe response to the invasion was to send I. and III./JG 53 across North Africa to Tunis, where both units were in action by mid-November. What the *Jagdflieger* found on this new front was hardly to their liking – the Allies had hundreds of fighters to protect their ground forces, many of these being new aircraft in the shape of P-38s, P-40s and Spitfires, all flown by USAAF units. If the enemy was not to make rapid gains, more German aircraft would have to be made available immediately.

The first *Gruppe* to arrive in response to this new crisis was

Take-offs and landings were always periods of excitement in a Bf 109 due to the aircraft's reputation for viscious ground loops. An event not to be missed, a fighter scrambling or recovering was usually watched by an audience on the ground, even if the occasion was only witnessed by a solitary mechanic, as in this photograph taken at a dusty Sicilian airfield in April 1943. This Bf 109G-5/Trop hails from II./JG 51 'Moelders', and wears the unit's distinctive buzzard's head emblem just forward of the supercharger intake

Oberfw Heinrich Bartels of IV./JG 27 in front of his Bf 109G-6/R6 'Kanonenboot', 'Red 13', which carried the nickname *Marga* on its port side. He scored his 70th kill in this potent fighter whilst based at Kalamaki, near Athens, with the remainder of this new *Gruppe* on 17 November 1943. Salient details like the yellow and white '87' Octane triangle and Directional Finding (DF) loop are clearly visible in this marvellous shot taken in Greece in late November 1943 *(Weal)*

The tail of Bartels' Bf 109 (Wr-Nr. 27 169) shows the airframe's spotless condition, and the carefully-applied victory bars. *Marga*'s pilot was the *Gruppe*'s leading *experte* at the time, and his flying garb is a little unusual for the theatre, particularly the black shirt. Bartels was lost following combat with P-47s over Bonn on 23 December 1944

This dramatic shot of a genuine *Alarmstart* for 4./JG 53 at La Marsa, Tunisia, in March 1943 shows 'White 4', flown by Oblt Fritz Dinger (67 kills), and 'White 5', the mount of Oberfw Stefan Litjens (38 kills), just seconds away from scrambling to intercept another Allied bombing raid *(Weal)*

II./JG 51, followed by II./JG 2 with Fw 190As – these were the first of the type to operate in the fighter role with any unit in the Mediterranean. Additional strength in Sicily was provided by Bf 109s of 11./JG 2, which then came under the operational control of II./JG 53. Support for II./JG 51 was also provided, by 11./JG 26.

In deteriorating weather the German pilots now found that they were additionally up against US heavy bombers, an all too familiar enemy to those who had flown in from Europe. One small ray of sunshine for the *Jagdwaffe* in all this seemingly endless gloom came in the form of novice US fighter units supporting Operation *Torch*. Ill-equipped for the deadly skies of North Africa, the USAAF squadrons had a costly introduction to combat at the hands of the *Jagdflieger*.

Also, there were occasions when the Germans came up against crews who may have been experienced, but were flying aircraft that were totally inferior to a Bf 109. The Bristol Bisley, a woefully inadequate derivative of the Blenheim V specially modified with a four-gun nose for close-air support work, was one such type, and when Oblt Julius Meimberg's 11.*Staffel* of JG 2 came upon a formation of 11 of them on 4 December, the *Jagdflieger* proceeded to shoot down every one – Meimberg himself claimed three. Such success was rarely repeated, however, for the Allies generally flew very capable aircraft, but the Bisleys (from No 18 Sqn RAF) had been part of 'maximum effort' operations to deny Bone, and other airfields, to the Germans, but it was foolhardy in the extreme to risk obsolete aircraft without an escort in an area still covered by Axis fighters.

JG 53's II.*Gruppe* now replaced III./JG 53, which was withdrawn temporarily. JG 77 remained in the south, with JG 2 and JG 51 in central Tunisia. Serious losses were now being felt by all *Gruppen* as the second front grew in size, these casualties being all the more serious when those pilots who failed to return were men who inspired others through their impressive scores. They were the cream of the *Jagdwaffe*, and virtually irreplaceable. Fw Anton Hafner (with 20 victories) was shot down and injured by Spitfires on 2 January 1943, although luckily for II./JG 51 he was hospitalised rather than killed. On the credit side, individuals such as Müncheberg, Bär, Hackl and Freytag continued to build on their already impressive tallies, the *Jagdwaffe* rarely shirking a fight even when the odds were highly stacked against them.

Among the '100 club' pilots operating in the south was Lt Wilhelm Crinius of I/JG 53, who added 14 kills in the Mediterranean to bring his total to 114, before he too was shot down by Spitfires on 13 January. Better luck attended Major Kurt 'Kuddel' Ubben, III/JG 77's Kommandeur, however, who achieved his 100th kill on the 14th. A highly successful *Jabo* pilot during the early days of the Mediterranean campaign (he, and another pilot,

had succeeded in crippling HMS *Warspite* off Crete on 22 May 1941 whilst fling Bf 109E-4s with JG 77!), Ubben subsequently became Kommodore of JG 2 in France. Also notching up kills was Lt Ernst-Wilhelm Reinert of II./JG 77, who had already scored 104 victories, most of which were achieved in Russia. He had an exceptionally successful day on 13 March 1943, claiming a P-40 over the Mareth line in the morning, followed by the destruction of no less than five P-39 Airacobras in the afternoon.

Having almost achieved a breakthrough at the Kasserine Pass, and given the US Army a bloody nose in the process, Rommel braced himself for a British thrust aimed at outflanking his defences at El Hamma. The front duly errupted into a series of bitter battles, and air activity was equally as intense overhead, with losses being experienced on both sides.

The Luftwaffe mourned the death of another fine pilot on 23 March when none other than Joachim Müncheberg failed to return from a *Frie Jagd*. It transpired that having come across a formation of USAAF Spitfires, Müncheberg quickly engaged them and damaged his 135th victim of the war. However, he had pressed home his attack a little too hard, and the great *experte* collided with the crippled fighter. Müncheberg's place was filled by Major Johannes Steinhoff, fresh from service with II./JG 52 in Russia. Boasting 150 kills, 'Macki' Steinhoff held strong views on Germany's prosecution of the war, and later wrote a balanced criticism of the high command's mis-handling of the *Jagdwaffe* in the Mediterranean. He was among the few who survived the debacle that was rapidly overtaking the German fighter force in the area.

Having achieved a tremendous total of close on 150 victories for the loss of 18 of its own aircraft, II./JG 2 returned with its Fw 190s to France in mid-March 1943. Sicily, in the meantime, had been reinforced against incursions by Allied bombers, II./JG 27 flying in to join III./JG 53. An additional, and extremely unpopular, job for the fighter pilots was to escort transports like the ubiquitous Ju 52 and giant Me 323 as they attempted to re-supply Rommel's forces in North Africa with fuel. The aerial route was now the only one left open as the Royal Navy had effectively blockaded the coastal ports. If Allied fighters intercepted the aircraft en route a large number of these heavily laden transports were doomed, as the *Jagdflieger* were only able to engage a few fighters at a time – the Allied pilots were also briefed to avoid the fighters and concentrate on the transports.

Sicily, and increasingly the Italian

More 4./JG 53 Bf 109G-4s at La Marsa, with 'White 7' and a chevron-marked machine on the extreme right-hand side of the photograph. All the Messerschmitts visible in this shot are fitted with 300-litre drop tanks, these external stores allowing them to patrol over the Tunisian frontline for longer periods. The tanks weren't overly liked by the *Jagdflieger* as they seriously compromised the Bf 109's manoeuvrability, and as soon as an interception was made they were jettisoned

That the Germans' Mediterranean war front had moved to more temperate areas by this stage was reflected in the mottled fuselages of Stab and II./JG 53's Bf 109G-2s, typified by this 5.*Staffel* machine photographed at Comiso in October 1942. The assorted paraphernalia of a unit on the move is visible in the foreground, the groundcrew busily ticking off the equipment list prior to loading everything aboard the Ju 52/3m parked in the background, and heading south for Tunisia

mainland, became the main staging areas for Luftwaffe units of all kinds, including fighters. Axis ground forces, being squeezed into a narrow area of Tunisia, were all but cut off from any assistance the Luftwaffe could provide.

On 20 April II./JG 51 fled North Africa for Sicily, its Bf 109s being pursued across the Mediterranena by Spitfires, which eventually managed to shoot down two of their number. Whilst all this was taking place, I.*Gruppe*'s Hptm Wolfgang Tonne got involved in a dogfight with a number of USAAF Spitfires over the Tunisian frontline, and after downing two for his 121st and 122nd victories (21 of these in-theatre), was killed when he crashed his damaged Bf 109G back at Bizerte.

Sorties from Sicily were now flown over those areas of Tunisia that still remained in Axis hands, thus offering pilots the advantage of crude airstrips where the short-range, drop tank-equipped, Bf 109s could quickly refuel and rearm mid-sortie. Despite the risk of their being captured (a fate which befell virtually all of them by late spring), the groundcrews continued their vital work to keep the fighters in action.

As part of a generally gloomy picture of the progress of the war in other theatres, the spring of 1943 soon recorded the last Allied push to defeat the remnants of the Afrika Korps in Tunisia. On 22 April this final offensive began, and it soon proved all but unstoppable. JG 77, operating from the Cape Bon peninsula, and JG 27 from Sicily, could offer only meagre resistance to enemy airpower that was stronger now than ever before.

By 7 May Bizerte and Tunis were in Allied hands, and remaining Luftwaffe units were ordered to evacuate to Sicilian or Italian airfields. Less than a week later, on 13 May 1943, North Africa was totally under Allied control. Mussolini, even with substantial help from Hitler, had failed in his bid to create a new empire in the region. The cost for the Luftwaffe in air and ground personnel alone was huge, with too many pilots and aircraft having been bled off for no gain whatsoever in a theatre which was never anywhere near as important to the Germans as Europe.

The striking double chevrons on *Gruppenkommandeur* Hptm Ernst Düllberg's III./JG 27 Bf 109G-4. More of a leader of men than an *experte*, Düllberg claimed only 10 kills whilst in-theatre. His distinctive rank patch is clearly visible on the right sleeve of his 'Fliegerblouse'

Another Bf 109 that changed sides was this G-2, the late property of the reconnaissance unit 2.(H)/14, whose little-known *Staffel* badge appears on the cowling. It was photographed at Gerbini on 5 September 1943, following the application of RAF roundels and the repainting of the upper surfaces and the fuselage in a darker shade of olive drab. Captured aircraft were rarely flown for long as they had a habit of going unserviceable very quickly due to a lack of spare parts

SICILY

It was more than clear that the Allies had no intention of making Tunisia the limit of their southern advance, and the Axis high command awaited the next blow with trepidation – wherever the enemy decided to strike, he now clearly held all the cards. Pilots of II./JG 27 might have anticipated the Allies capturing Pantelleria, a heavily fortified island lying midway between the African coast and the Italian mainland. Instead, it was decided (as something of an experiment) to recapture Pantelleria by massive air bombardment alone. This decision put hundreds of medium bombers over the island in successive waves, and brought the *Gruppe* considerable combat

success, with 25 kills being scored between 18 and 31 May.

Sicily also had to be defended, for this was most likely to be where the next major Allied operation was going to take place. While II./JG 51 and a substantial Italian fighter force remained in Sardinia, the main *Jagdwaffe* element defended Sicily, the island now being strengthened by a number of *Schlachtgruppen* equipped with Fw 190s. The Bf 109 continued to predominate in the interceptor role, however, the force being equipped with the G-6 and G-10 variants – both were fitted with heavy cannon and rockets to supplement their integral weaponry.

Fresh aircraft were being continually issued to III./JG 27, and a new IV. *Gruppe* was then in the process of being formed. JG 77, meanwhile, had to make do for the time being without the majority of its groundcrew, who had had to be left behind when Tunisia was abandoned as a base for fighter operations. Stab, I. and II. *Gruppen* remained on Sicily, while JG 77's III. *Gruppe* went to join II./JG 51 in Sardinia. In July IV./JG 3, a new unit, also arrived at Foggia, led by Hptm Franz Beyer, a Russian Front veteran with JG 3 who had shot down over 70 aircraft.

Few doubts remained that Sicily was next on the Allied invasion timetable, and after a swingeing piece of verbal abuse from Goering himself following the North Africa debacle, the *Jagdflieger* stood ready to face the onslaught. Even if some pilots now questioned what value their continued sacrifice was to the Reich, there was little talk that could be interpreted as defeatist.

Adolf Galland was present in Sicily at this time in an attempt to bring all available *Jagdflieger* together. His idea was to strike a crippling blow to the US heavy bombers then flying freely over the island. If the Bf 109s could shoot down a number of bombers from one raid, Galland believed that the invasion might be delayed, even if only for a short time. His plan utterly failed, however. Caught between a B-17 raid and a low-level strike by B-26s, the Bf 109s didn't manage to intercept either force. Only 'Macki' Steinhoff made contact with the heavy bombers, and duly shot down a single Fortress. The telegram from Goering that followed threatened one pilot from each of the *Jagdgruppen* present with a court martial!

Bf 109G/R6 'Kanonenboot' 'Yellow 14' of 6./JG 53 comes under close scrutiny from an official RAF Inspection team soon after being acquired following the capture of the large Sicilian airfield at Comiso in the late summer of 1943. The barrel of the starboard MG 151 cannon appears to have already been souvenired by a member of the 8th Army! Parked in the distance behind this machine is a Spitfire Mk VC from one of the first Allied fighter units to fly into Sicily

This ridiculous threat was never carried out, but its effect on morale was hardly calculated to improve a bad situation. It was not that the *Jagdflieger* were incapable of inflicting heavy casualties, as numerous American bomber crews could attest to, but the lack of numbers now worked against them most of the time – not always, however. On 2 July, II./JG 27 shot down four B-24s over Lecce, and the following day I./JG 77 despatched five P-40s while the *Geschwader*'s II. *Gruppe* was hunting B-17s and a reconnaissance Mosquito. The latter was finally intercepted and shot down, thus becoming the first of its type to be lost in this war zone. PR overflights were a continual annoyance, and every effort was made to catch them, although usually with little success.

SICILY INVADED

Despite disrupting several PR flights, Allied intelligence on Luftwaffe strength was more than adequate, and on 10 July 1943 Operation *Husky* began. In little more than a token gesture of defiance, the fighters in Sardinia were recalled, 39 Bf 109s of II./JG 51 responding immediately. This understrength *Gruppe* co-ordinated operations with Stab and II./JG 27, and II./JG 77 at Trapani, with all component *Staffeln* seeing action from the start of this new crisis. Pilots could not complain about a lack of targets, for Allied aircraft swarmed all over the island looking for the Luftwaffe.

Against such hordes the *Jagdflieger* was virtually committing mass suicide even in attempting to take-off, and on the opening day of *Husky*, claims of six bombers had to be offset by the loss of four Bf 109s. Gerbini became a nightmare of overlapping bomb craters as the *Jagdwaffe* strove to carry out the dual task of both blunting enemy bombing and covering the Straits of Messina, the strip of water which represented an escape route to Italy. Makeshift sites away from the bomb-blasted airfields enabled sorties to continue, but by 13 July all units except II./JG 51 had moved back to north-eastern Sicily. The withdrawal to Italy, which saw survivors head for the complex of bases on the Foggia plain, now began for the handful of aircraft that had been left intact by Allied bombers.

Some pilots managed to beat the odds, including Werner Schroer. Reinforcements in the shape of the Bf 109s of 8./JG 27 flew to Brindisi, Italy, on 15 July. Their *Staffelkapitan* was Oblt Wolf Ettel, whose 120 victories were rapidly added to in the fight for Sicily when he shot down five aircraft in three days. Combining with II./JG 27 on the 16th, the Bf 109s attacked a formation of B-24s and claimed nine, two each falling to Ettel and Schroer. Ill-fortune struck the following day, however, when 8.*Staffel* lost four killed. The victims included Ettel, who was carrying out a low-level attack south of Catania when he was hit by AA fire.

II./JG 51 now also joined the general withdrawal to souhern Italy, the *Gruppe* having been reduced to just four aircraft. Most of the rest had been disabled or destroyed on the ground by bombing. Despite overwhelming odds, the Wehrmacht was not yet a totally spent force in Sicily, and when the embattled troops needed supplies on 25 July, JG 27 and JG 77 desperately tried to

A photograph that exemplifies the Luftwaffe's fate on airfields all over the Mediterranean as the Allies advanced under a huge umbrella of airpower. Bombed on their landing grounds, the *Jagdwaffe* failed to keep pace with the appalling attrition rate. These JG 51 G-5/Trop wrecks had been bulldozed to one side by Royal Enginneers prior to declaring Comiso open for Allied squadrons to fly in. The machine on the right wears the chevron marking of a *Geschwader* adjutant *(Robertson)*

Surrounded by a familiar cloud of dust, a Bf 109G-6 of II./JG 51 taxies out for take-off from a Sardinian airfield in the fateful summer of 1943. This unit's style of presenting the II.*Gruppe* horizontal bar forward of the number, rather than aft of the *Balkenkreuz*, was unique

Fine detail view of the 'Kanonenboot's' twin 20 mm guns that gave the Bf 109G-6/R6 a very useful increase in firepower. Still fitted with its belly tank, this JG 53 fighter was photographed returning to its Tunisian base following an uneventful sortie in early 1943 – a rare event at this point in the campaign

shield a formation of Ju 52s from marauding Spitfires, but to little avail. Along with the transports lost was the Bf 109 of Oblt Heinz-Edgar Berres, a 53-victory *Ritterkreuztrager* of I./JG 77. Another *experte* killed at this time was II./JG 53's Oblt Fritz Dinger (67 kills), who was mortally wounded on the ground by bomb fragments during an attack on his temporary airfield on Sicily on 28 July. Three days later , II./JG 27 was ordered back to Germany for Reich defence duty, its Bf 109s being left behind to be distributed throughout JGs 3, 53 and 77 at Foggia. II./JG 51 also returned home for re-equipment.

Italy itself looked increasingly like a lost cause to the German high command. The defence of Germany itself demanded an even greater effort from the *Jagdwaffe* as Allied bombing raids were growing ever-more intense in the south. Neither was Italy immune from attack, the Allies sending missions against industrial areas, railways and Luftwaffe bases.

Allied fighters were also based close enough to carry out damaging strafing attacks on German bases in Italy, although these, hitherto carried out with low casualties, were now hotly contested as the Luftwaffe sent more fighters into the region. For example, an attack on the Foggia complex by P-40s on 25 August was hardly molested, the fighters destroying mostly Axis bombers. Five days later, despite the medium bombers only suffering light casualties, their P-38 escort lost 13 to Bf 109s. But just as the Luftwaffe could afford to lose all but impotent bombers on the ground, so it yielded the *Jagdwaffe* little gain to shoot down US fighters. Naturally enough, neither side could always pick the most ideal target.

SALERNO

On 1 September Allied armies went ashore on the toe of Italy virtually unopposed by the Luftwaffe. However, on the second day the *Jagdwaffe* retaliated, and in attempting to destroy bombers attacking the marshalling yards at Cancello, instead found P-38s in their sights. Ten Lightnings went down, but this score was not achieved without cost for II./JG 53 lost the 53-victory *experte* Oblt Franz Schiess, whose aircraft crashed into the Mediterranean. Then came a body blow that many Germans never expected. Italy surrendered to the Allies and Adolf Hitler was faced with either abandoning the southern front entirely, or fighting on irrespective of what his former Allies had decided. He chose the latter course, there being little breathing space before a stand had to be taken against the invasion force pushing rapidly inland from Salerno.

Foggia was abandoned at the end of September and the *Jagdwaffe* headed north to airfields around Rome. IV./JG 3 returned home from Sardinia, which was evacuated to save casualties. III./JG 77 left for Rumania and II./JG 53 for Austria. I./JG 51 returned to southern Germany and Hptm Hans 'Gockel' Hahn's I./JG 4 flew into northern Italy from Rumania.

ITALIAN DEBACLE

In late January 1944 the Allies put troops ashore at Anzio with the aim of forcing a route through to Rome, as well as knocking out the German forces defending the difficult mountainous terrain dominated by the monastry of Monte Cassino. Initially the landings went ahead virtually unopposed, but the Allies' hesitancy in breaking out from the beach-head almost turned the invasion sour, as Kesselring poured in troops and panzers to pose a dangerous threat to the invaders.

In what was now a familiar pattern, the *Jagdwaffe* did its upmost to support the ground forces which were, for the first time in months, moving forward rather than retreating. But Allied fighter-bombers ranging out from southern Italian airfields made Bf 109 sorties increasingly hazardous. Striking northwards, American fighters and medium bombers pinned down the renmants of the *Jagdwaffe*, which had also to contend with increasing pressure from the bomber fleets of the Fifteenth Air Force.

II./JG 53 left these Bf 109Gs behind at Comiso during the frantic evacuation of Sicily, their 'patch' of airfield being shared with a shattered Ju 88A-4 that exhibits the distinctive *Wellenmuster* (wave pattern) camouflage worn by anti-shipping bombers on Sicily. A 'cannonless' G-6 Stab aircraft, wearing a familiar interceptor striped propeller hub, is on the left, and a G-6/R6 still boasting its MG 151s is on the right *(Weal)*

During October 1943 Oberst Baron Günther von Maltzahn, Kommodore of JG 53, was appointed to the new post of *Jagdführer (Jafu) Oberitalien* in command of all fighter units in Italy. A Stab had in fact been established as early as July that year, with Oberst Alexander Lohr as Chief of Staff finishing an eight-month (January to August 1943) tenure as overall Lufwaffe commander in Italy as the German commitment wound down.

It appears that the new Chief of Staff was Wolfram von Richtofen, who was replaced around August 1944 by Gen Ritter von Pohl, formerly the German air attaché in Rome. As the tactical commander, Oberst 'Edu' Neumann was eventually succeeded by Oberst Gunther 'Franzl' Lutzow, the latter being posted to what was then considered by the Germans to be a mere backwater of the war following his contretemps with Goering in the so-called 'plot' against the Reichsmarschal in January 1945. Lutzow's new command had in fact ceased to exist, *Jafu Oberitalien* having been disbanded in December 1944 and apparantly not replaced, at least by any German command. Lutzow is believed, therefore, to have acted as a liaison officer to the ANR for the remainder of hostilities.

Obliged to oppose the heavy bombers of the Fifteenth USAAF, the Axis had also to contend with the fighters on Corsica; P-47 Thunderbolt units based there were increasingly effective in interdicting the Italian road and rail system to deny the Germans supplies coming down north of Rome. Medium bombers were also much in evidence, the entire Allied air forces ostensibly offering the Germans and Italians targets flying at low, medium and high altitudes. The weary *Jagdflieger* now knew that the air war over Italy was all but lost.

Tasked with a consequent diversity of effort gave little respite to the thin resources of the fighter force, although the Luftwaffe made an attempt to ensure that units should counter each of the threats independently. Thus, JG 77's duty became primarily air defence while JGs 4, 51 and 53 flew support to the ground operations at Anzio and Cassino. Many sorties were also intended to ward off enemy fighters so that the *Schlachtgruppen* could get through to their targets. At times it was all but impossible to prevent the enemy from shooting down the Fw 190 *Jabos* in droves. That the opposition was overwhelming was shown by the record of I./JG 4, who lost 14 pilots killed, wounded or captured in the first six weeks of 1944. The casualties included its Kommandeur, 'Gockel' Hahn, who was lost on 27 January.

On 3 February II./JG 51's Hptm Herbert 'Puschi' Puschmann was killed during an interception of a B-25 formation. His rear-quarter attack was met by heavy fire from the Mitchells' tail gunners, which sent the Messerschmitt down in flames near Civitavecchia. Another *Ritterkreuztrager* lost in action in this theatre, Puschmann had scored a total of 54 victories at the time of his death. February 1944 also recorded the final reinforcement of the *Jagdwaffe* on this front when I./JG 2 brought its Fw 190s in from France. Much of the interception work now concentrated on the US heavy bomber en route to their targets in Germany. Over Italy and the Balkans, air combats brought the *Jagdflieger* limited success against their primary targets as the ratio of Allied escort fighters to bombers steadily increased, often exceeding two to one.

With the focus now firmly fixed on other areas, the future defence of

A mixed formation of Bf 109G-5/6s of 7./JG 27 is seen on a bomber escort mission over the Adriatic from their base at Kalamaki, in Greece, in January 1944. Leading the force in his G-6/R6 Trop, coded 'White 9', is the *Staffelkapitan* himself, Oblt Emil Clade. Already an *experte* when this shot was taken on an overcast winter's day, Clade, and his *Staffel*, were posted back to Vienna just weeks later. He somehow managed to survive the Defence of the Reich bloodshed that was III./JG 27's lot for the remainder and war, ending the conflict with 26 kills to his credit *(Weal)*

A close-up shot of the same formation reveals the 'Kannonenboot' configuration of 'White 7' and '9'. 'White 2' is a late-build G-5/Trop, and like its *Staffel* mates, appears to be in almost factory-fresh condition *(Weal)*

Italy looked precarious at best. Crete still remained in German hands, however, despite the island being regularly bombed by RAF squadrons based in Egypt. A small detachment of II./JG 27 was left behind to provide a token defence, and on 6 March 1944 a *Staffel* of Bf 109Gs intercepted a force of No 3 Wing SAAF B-26s and destroyed four of them, although this was their only real success. Several other 'blockading' missions were mounted on Crete, but the end result of these engagements was an

inconclusive trading of fire which only damaged aircraft on both sides.

More personnel changes were made at this time, and able pilots who could make a more worthwhile contribution to the disasterous situation over the Reich left Italy for the last time. Added to combat losses, these postings served to denude even further the already-meagre pilot strength of the remaining Italian *Gruppen*. They were simply being swamped, and as the number of replacements for casualties dropped to a trickle, the Luftwaffe High Command made it clear that no further units would be sent – indeed, there were precious few that could now be spared. Faced with this situation, there was little the *Jagdwaffe* could do to stop the Allied spring offensive of 1944 breaking through the Gustav line of fortifications and linking up with the forces at last breaking out of from Anzio.

By the time Rome fell to the Allies on 5 June 1944, only JG 77 and I./JG 4 remained in Italy. To cap it all, the invasion of Normandy took place the very next day, thus opening another far more dangerous front to suck in available German fighter forces. There was also a Russian summer offensive for the Germans to contend with – the culmination of all this defensive posturing on all fronts saw Italy totally abandoned by the *Jagdwaffe*. By the end of June 1944, JG 77 had left the country, along with I./JG 4, thus ending the Luftwaffe's brief and bloody struggle against overwhelming odds in this theatre of war.

A NEW AIR FORCE

Considerable confusion had arisen following the Italian surrender; many military personnel, particularly pilots, felt totally let down by the ignominious cessation of hostilities, which in their eyes amounted simply to defeat. The release of *Il Duce* from a mountain prison in September 1943 revitalised those sections of the armed forces who still felt an allegiance to the fascist regime, and on 10 October Mussolini declared the formation of a tri-service Aeronautica Nazionale

A curiously attired Axis soldier peers into the empty cockpit of 7.*Staffel* G-6/R6 Trop Wr-Nr. 15 508, that had been temporarily abandoned where it stopped following a heavy landing at Malemes in north western Crete on 1 December 1943. The distinctive 'pierced apple' *Staffel* badge under the cockpit and the III.*Gruppe* emblem on the cowling are both clearly visible in the shot. This isolated outfit were very diligent in maintaining their *Gruppe* identity, virtually every Bf 109 on strength wearing this elaborate marking *(Weal)*

Spiral spinner decoration reflected the increasing demand for the Bf 109 units in the Mediterranean to undertake high-level bomber interception duties, a task for which the Messerschmitt fighter remained lethally suited. This G-5/Trop of I./JG 53 boasts a complete set of the correct Mediterranean theatre markings for this late stage of the campaign

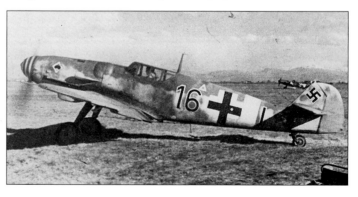

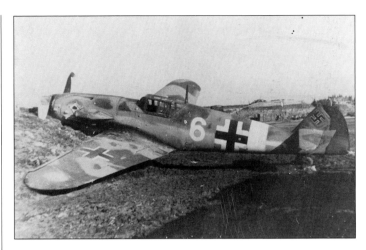

All fighter units tended to have aircraft with colour schemes that were adapted to local terrain, and this Bf 109G-2 of II./JG 53, seen in mid-1943, appears to have been entirely repainted at some stage – only to be damaged by the pilot losing control on a bumpy taxyway and running smack into an unforgiving earth wall. Although the damage appears to be restricted to nothing more than a bent propeller, a shortage of even basic spares like airscrew blades in this late stage of the war probably resulted in this otherwise serviceable fighter being abandoned where it came to rest

Republicana, or simply the ANR.

In the event it was the air force which was to see almost all remaining action by Italian forces against the Allies, and to this end pilots sympathetic to the cause were organised into units at airfields in the north of the country. Reporting centres were established on 15 October for dissemination of crews into the various types of unit, Milan's Bresso airport becoming that for fighter pilots.

101° *Gruppo Autonomo Caccia Terrestre* was probably the first unit to form in Florence late in 1943. Pilots transferred to Mirafiori, near Turin, in 1944 and subsequently went to Germany to convert to the Bf 109G. I° *Gruppo Caccia* officially formed on 1 January 1944, and was equipped with Macchi C.205 Veltros, which had been used by II./JG 77 for a short period following the armistice with German markings applied directly over their Regia Aeronautica camouflage. This strong German link remained with the new unit as pilots transferred to Lagnasco for instruction in Luftwaffe fighter tactics – the Macchis, meanwhile, had been returned to Italian control. The other unit formed on 1 January was the so-called *Squadriglia Complementare 'Montefusco'*, who flew a mix of C.205s and Fiat G.55s. Both these types represented the peak of Italian fighter develoment, and possessed excellent all-round performance.

ANR units were formed primarily for local defence, there being little desire on the part of the pilots to attack the southern, Allied-held, areas of Italy and thereby risk combat with aircraft of the Italian Co-Belligerent Air Force. Furthermore, Italy's industrial heartland lay almost entirely in the north, and that would be the area where Allied air attacks would most likely occur. A ready supply of new aircraft was therefore available, in addition to those examples handed back by the Germans who had requisitioned most military equipment immediately the armistice had been announced.

Realising that the ANR's existence could be brief, and their position all but hopeless, did not deter the Italian fighter pilots who fought, much as the Germans, for far more basic ideals than those postulated by politicians. Operations began on 3 January when ten I° *Gruppo Caccia* Macchi C.205s scrambled to intercept a Fifteenth AAF B-17 formation escorted by P-38s. Three victories were claimed.

Personnel of JGs 2 and 77 worked with their Italian opposite numbers to direct and control fighter operations at Rivolto and Senago, the two *Gruppen* collaborating with I° and II° *Gruppo* respectively. Operating alongside JG 77, I° *Gruppo* moved to Campoformido, near Udine, to be deployed almost exclusively against the Fifteenth's heavy bombers. By combining the German and ANR units, the Axis force could regularly put up nearly 100 aircraft on successive operations. Air combat victories tended to be confined to small numbers of bombers, however, and the ever-present US fighter escort posed a great threat.

Italian officers examine the scoreboard on the tail of Haupt Werner Schroer's Bf 109G-2/Trop on the Greek island of Rhodes in the spring of 1943. He was 8./JG 27's *Staffelkapitan* at the time, and his score then stood at 60. Completing several other frontline postings following his long spell in the Mediterranean, Schroer eventually ended up as *Kommodore* of JG 3 'Udet', defending Germany. One of the luckiest of all veteran *Jagdflieger*, Schroer finished the war with 114 kills to his credit, including no less than 26 four-engined bombers – no wonder his Italian guests look suitably impressed!

II° *Gruppo Caccia* was formed at Bresso during the spring, this second ANR fighter unit comprising a nucleus of ex-3 and 150 *Stormo Automomo* pilots who had flown the Bf 109G for the last three months as part of the Regia Aeronautica. Early in March training began on the Fiat G.55. That month saw considerable action by I° *Gruppo*, with four B-24s and eight P-47s being claimed against the loss of four C.205s. Inevitably, it was not long before the ANR's bases were bombed, and on 18 March Luftwaffe Bf 109Gs and C.205s intercepted a Liberator force briefed to knock-out airfields in the Fruil area. Four B-24s and three of the P-38 escort were claimed for two C.205s shot down. The bombing was relatively ineffective though, with only two ANR fighters being destroyed and five damaged on the ground.

I° *Gruppo* saw more action before the end of March, claiming ten enemy aircraft destroyed for two fighters lost. On the 29th all three Italian fighter units scrambled to take on a force of P-38s and claimed two Lightnings for the loss of two C.205s. Five P-38s were believed damaged.

Gruppo 'Montefusco' also did well, its G.55s shooting down two B-24s from a force briefed to bomb airfields in the Milan area, with a third claimed as a 'probable' and a fourth forced to crashland damaged, but intact, at Venegono. Unfortunately, the unit's CO, Capitano Bonet, was lost in one of the two G.55s shot down. II° *Gruppo* also made its combat debut during this action, but its G.55s reported no victories or losses.

Among the problems the ANR fighter units had to contend with was the superficial resemblence of both the C.205 and G.55 to the P-51 Mustang. Even in mixed formations mistakes could occur when combat was joined, and JG 77 pilots occasionally mistook their allies for the enemy, with disastrous results. For example, on 29 April the Bf 109s shot down two C.205s, both pilots being lost.

II° *Gruppo* made its combat debut on 30 April, claiming one B-24 for one fighter. Action was infrequent and, sometimes, too close to home. Concerned primarily with escorting tactical bombers, Allied fighter units found little opportunity for air combat against an enemy whose appearances were becoming increasingly rare. On 27 May, however, the Spitfire Mk IX pilots of No 1 Sqn SAAF engaged a different kind of target following a sweep against the Italian airfields at Rieti, Terni and Foligno. Axis fighters reacted immediately, and six Spitfires came under attack from what were reported at the time to be Fw 190Ds, although it appears more likely that these were C. 205s. One of the 'Focke-Wulfs' was shot down and its pilot baled out, and he was immediately joined by a second.

It was at that point that Lt C Boyd, 'recording the presence of the Fascist Republican Air Force for the first time in contact with the SAAF', chased a 'Bf 109 with Italian markings'. This was actually a I° *Gruppo* C.205, and he soon shot it down. A second Macchi that was attempting to land at one of Foligno's satellite fields blew up shortly afterwards following several accurate bursts of fire from Spitfires flown by Lt Col Bosman and Lt T E Wallace.

Bosman also accounted for another Bf 109 later in the mission, making the unit's score for the sortie four destroyed in aerial combat, plus one on the ground, and two or more damaged. The SAAF subsequently picked up confirmation when they overran I° *Gruppo*'s HQ at Reggio nell' Emilia (the unit having transferred in from Veneto), that Capitano Sergio Giacomello and Sergente-Maggiore Giorgio Leone had been lost in this action. The headquarters of I° *Gruppo* had itself been attacked on 12 May when P-38s attacked Reggio nell' Emilia airfield, their strafing runs destroying two C.205s and damaging six.

In order to reduce the recognition problem (neither side was in a position to waste aircraft, let alone pilots) Luftwaffe High Command wisely decided to re-equip the ANR units with late-build Bf 109Gs. The Germans were very pleased with the performance of the Italians who, despite having had relatively few opportunities to demonstrate their skills, had hardly been overwhelmed by Allied incursions over their territory. However, having entered the fray in only modest numbers from the start, the continuous spring combats soon forced I° *Gruppo* to stand down, and found II° *Gruppo* suffering from a shortage of spares. Training using Bf 109G-6s passed on by I./JG 53 and II./JG 77 began in June for those ANR pilots who had not flown the type previously.

I° *Gruppo* returned to the fray in June 1944, making nine scrambles between the 4th and the 20th in G.55s and C.205s passed onto them by II° *Gruppo*. Four days later II° Gruppo took its Bf 109s into action for the first time and shot down two P-47s without loss. While awaiting a full complement of Messerschmitts, II° *Gruppo* maintained a single *Squadriglia* on the G.55 until July.

Better results were obtained with

A very youthful looking *Jagdflieger* from Schroer's *Staffel* explains the intricacies of flying the G-2/Trop in combat to a rivetted audience of German and Italian aviators. The varied assortment of clothing worn by the pilots in this shot typifies the relaxed attitude to flying and personal kit that existed in the Mediterranean throughout this bitter campaign. Relations between flying and ground personnel of the Axis powers was generally cordial, although the Italian authorities did not overly encourage fraternisation the Bf 109, although overall the Italians' claims still remained modest. In seventeen July combats ten Bf 109s were lost for ten A-20s, six P-47s, four Spitfires, three B-24s and a P-38 destroyed. In August III° *Gruppo* was formed, it being intended that I° *Gruppo* should take over JG 77's remaining aircraft once the latter had returned to Germany. At that point, however, a crisis occurred that threatened to wreck the links forged between the Luftwaffe and the ANR.

The Luftwaffe's *Italien* command, which directed ANR fighter operations, now began moves to bring the Italian units totally under German control. An attempt to persuade all personnel to join the Luftwaffe failed, and the Germans then foolishly attempted to force the issue. In August 1944 the ill-starred Operation *Phoenix* began, with Luftwaffe officers being sent to all ANR units to inform the Italians that their autonomous force had been disbanded. Instead, they were to be formed into an 'Italian Legion', or posted to Flak units.

Reaction to this political move varied; with few deeply held nationalist convictions left, many Italian pilots simply ignored the dictates, while others were on the point of violent rebellion against their erstwhile ally. Mussolini protested to Hitler, and von Richtofen and his staff were recalled to Germany to be replaced by Gen von Pohl.

During the two months that *Phoenix* was in place the ANR units were paralysed, and it was only then that the worth of the 'air force within an air force' was truly realised by Luftwaffe High Command. There was little doubt that the Italians could be counted upon, and their presence and willingness to continue the fight meant that the last Luftwaffe units could now return to Germany.

The end result of this fiasco was that I° and III° *Gruppo* received the Bf 109G immediately, and while they were away in Germany for their conversion courses, II° *Gruppo* would remain as the sole fighter unit defending all of northern Italy after the last of the *Jagdwaffe Gruppen* went home. Whilst all this re-organisation was taking place, the *Squadriglio* still on the frontline found little shortage of targets, and on 19 October II° *Gruppo*'s Bf 109s intercepted USAAF B-26s and claimed eight shot down for one of its own. Three more kills were scored before the end of that month.

II° *Gruppo*'s activities from its new base at Aviano were becoming increasingly annoying to the Allied Strategic Air Force command which had, according to the Italians, lost seven B-17s, five B-26s, two P-47s and a P-51 in return for four Bf 109s by 15 November. On the following day the ANR flew their largest operation of the war, intercepting Fifteenth AAF bombers in some strength – the USAAF crews reported seeing up to 40 enemy fighters during the engagement. Despite the efforts of the P-51 escort, 14 'heavies' were reported missing. Mustangs destroyed eight of the attacking fighters, plus two 'probables' and two damaged. In addition the 'heavy' bomber gunners claimed to

Although the exact unit allegiance of these Italian pilots cannot be confirmed, it is likely that they were from one of the *Squadriglio* within 161° *Gruppo*, who flew Macchi C.202s from a nearby airfield on Rhodes. If Italy had stayed in the conflict past September 1943 it is likely that these Regia Aeronautica pilots would have found themselves flying Bf 109Gs from their Greek island base. Both German pilots in the foreground are wearing light-weight life-preservers over their tunics – a vital piece of personal equipment for over-water operations

The sheer waste of machinery by the Luftwaffe in the last 18 months of the war in the Mediterranean is graphically illustrated by this shot, taken only a matter of days before the final evacuation of Tunisia in May 1943. The one-and-a-third Bf 109G-2/Trops belonged to I./JG 77, this unit subsequently having the distinction of being the last frontline *Geschwader* in Italy. A crippled Ju 52/3M sits where its undercarriage collapsed, whilst other serviceable transports fly out the handful of remaining groundcrews *(Weal)*

Exemplifying the close co-operation achieved by the Luftwaffe and ANR is this view of Bf 109G-6 'White 3' of 4./JG 77 and a Republican Macchi C.205, taken over the northern Italian Alps during mid-1944. Very few photographs were taken of ANR and *Jagdwaffe* fighters together either on the ground or in the air, a fact which makes this unusual shot all the more rare. Like their German partners, the Italians have applied spiral markings to the spinner of this C.205, thus denoting its bomber interceptor role *(Weal)*

have shot down a single fighter.

To wipe out this last-ditch enemy defence, a series of heavy raids were accordingly made against ANR airfields at Aviano, Vicenza, Villafranca and Udine. SAAF medium bombers attacked these four bases on the night of 17/18 November, and followed up with a day raid on the 18th. No less than 186 P-51s patrolled the target areas to deal with any opposition. Although only a handful of Bf 109s were destroyed on the ground, the Italians now faced a predicament similar to that of the Germans in that aircraft could be replaced easily, but pilots killed or injured in such raids could not. I° and III° *Gruppo*, meanwhile, continued their training while II° *Gruppo* 'held the line' through the winter. A shortage of both fuel and spares bit hard at the frontline, and it came as welcome news for the battle-weary II° *Gruppo* that I° *Gruppo* had returned to Italy in January with a full complement of Bf 109G-10s. III° *Gruppo* remained at Fürth for the time being.

Maggiore Miani commanded II° *Gruppo* throughout this period, and the unit got results on four February missions, with a quartet of Bf 109s being lost for the destruction of ten B-25s and a Spitfire. The US medium bombers were busy at this time pounding the remaining Italian rail links to the rest of Europe, with a large number of sorties being flown against the Brenner Pass and other key areas. The ANR's last recorded success was an attack on B-26s of No 3 Wing SAAF, and although claims for eight bombers were filed (for the loss of two Bf 109s), no such substantial loss appears in SAAF records. This leads one to believe that these Marauders, if indeed that is what they were, came from another air force.

There was further sporadic contact with Allied fighters but the clashes were rarely productive for the defenders. Any hint of unescorted bombers being attacked by enemy fighters simply brought swift and powerful response from the massive Allied escort force in-theatre – a daunting prospect for the Italian pilots.

During its short operational life, the ANR had generally flown its missions with skill and courage, particularly when one considers the heavy odds stacked against it. Combined air combat victory claims for the three main *Gruppo,* plus the *Squadriglia Caccia Complementare,* were 226 aircraft destroyed for the loss of 137 fighters of the three types flown. I° *Gruppo*'s Maggiore Adriano Visconti scored seven victoiries with the ANR, to add to his 19 from his Regia Aeronautica days, whilst II° *Gruppo*'s two leading pilots were Tenente Ugo Drago and Capitano Mario Bellagambi, who both finished with 11 victories each.

Concrete proof of the esteem in which the Italians were held by the Germans was revealed in a plan uncovered at the end of the war which would have seen ANR pilots trained as the first foreign nationals to fly the Me 262. Two airfields had been designated for jet operations, but the final Axis collapse came before these plans could be implemented, and apart from a few isolated skirmishes with US fighters, there was little further action for the ANR.

JAGDGESCHWADER 27 'AFRIKA'

Considering *Jagdgeschwader* 27's future service in North Africa, the fact that I.*Gruppe* adopted a badge depicting a stylised leopard threatening a startled Negress, superimposed on a map of Africa, was remarkably prophetic. The badge was applied to Bf 109Es flown by I./JG 27 as early as the spring of 1940, when it was based at Etaples in the Pas de Calais for operations across the Channel.

A JG 27 Stab was first raised in 1939 and used to form that of JG 77, a new Stab/JG 27 being stood up at Krefeld on 1 October 1939 when I.*Gruppe* also formed with the standard three *Staffeln*. Oberst Max Ibel was Kommandeur of the Stab, whilst I.*Gruppe* was led by Hptm Adolf Galland. A II.*Gruppe* Stab was organised in January 1940, although the component 4., 5. and 6.*Staffeln* did not adopt the *Geschwader*'s numerical designation until July of that year. The 'Sitzkreig' war peroid was marked by training and patrols along the Franco-German border until 10 May 1940 when Germany invaded France and the Low Countries. I.*Gruppe* was then under the command of Major Eduard Neumann, and whilst flying support missions for Operation *Weserubung*, scored its first victory during this initial week of fighting – tasked with helping subdue

Oblt Erbo Graf von Kageneck, *Staffelkapitan* of 9./JG 27, was photographed in May 1941 on Sicily, strapped into his Bf 109E-4 and in the process of running through the all-important fight control checks with his trusty 'black shirt', prior to departing on a sortie over Malta. Note the metal rank pennant attached to the radio mast, and the unusual positioning of the aircraft number on the engine cowling – all I./JG 27 fighters appear to have had these individual *Staffel* numbers applied in-theatre at about the same time as the all-yellow Balkan campaign nose marking was sprayed on. Having scored 65 kills by December 1941, which made him the *Geschwader*'s top scorer, Erbo Graf von Kageneck was killed in action on Christmas Eve

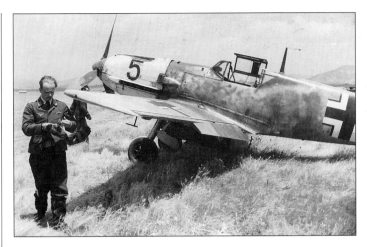

the Belgian Air Force, the *Gruppe's* Heino Becher destroyed a solitary Gladiator near Tirlemont in this brief operation.

Covering the 6th Army's drive using airfields at München-Gladbach and Gymnich, near Cologne, I./ JG 27 had I./JG 1 and I./JG 21 under its operational control throughout this lightning campaign. These units had a memorable 12 May, flying 340 sorties and destroying 28 enemy aircraft for the loss of four Bf 109Es.

III./JG 27 also formed in July 1940 with its component 7., 8. and 9.*Staffeln*, all of which were created from a nucleus of personnel from other fighter units as was standard Luftwaffe practice. In their turn, as attrition all but removed the entire original air component, new 'replacement' *Staffeln* were raised, these taking the same numerical designation.

In May 1940 *Stab* IV./JG 27 was formed, with 10., 11. and 12.*Staffeln* being organised at the same time. Additional *Staffeln* (13., 15., 14., 15. and 16.) were added in November of that year. In addition, JG 27 had an *Erg* (Training) *Staffel* from November 1940, this later (in February 1941) being expanded to *Gruppe* strength.

Having fought in the Battle of France, JG 27 suffered substantial losses, along with other units of the *Jagdwaffe*, whilst trying to subdue RAF Fighter Command during the Battle of Britain. For a combined victory total of 146, the *Geschwader* lost 56 pilots killed or missing, including the Kommandeur of

The island garrison of Malta was attacked by both fighter and *Jabo Staffeln*, the latter being deployed by 8./JG 27. Bf 109E-7 'Black 5' of this specialist fighter/bomber unit is seen here in May 1941 fully bombed up with a single SC250 device fitted to the centreline SC500 stores rack

An armourer strains to roll an SC250 bomb across an overgrown dispersal at a Sicilian airfield in late May 1941. Out of shot to the left of the photograph is another similarly struggling groundcrewman, endeavouring to pull the bomb over to an as yet unarmed Bf 109 *Jabo*

Greatly assisted by the Luftwaffe's ubiquitous hand-operated bomb hoist, armourers prepare to lift the weapon up to its mounting. As with *Jabo* operations elsewhere, JG 27's effort was on too small a scale to make a great difference to the ill-fated assault on Malta

71

III. *Gruppe,* Hptm Joachim Schlichting, who was seriously injured after being shot down by Spitfires on 6 September.

One small compensation was that by the end of 1940, JG 27's surviving pilots had acquired considerable experience of British tactics. While the assault on England had been a sobering, tough, lesson in flawed strategy, *Jagdwaffe* fighter tactics had often proved basically superior to those of the RAF. Such knowledge was to stand the 'old heads' in good stead when *Gruppe* transferred to North Africa in April 1941.

Despite having to fly obsolescent Bf 109E-7/Trops for the early months of combat in the desert, the veteran pilots of the *Geschwader* experienced few problems in dealing with the Allied fighters in-theatre. By June 1941, however, the first examples of the Bf 109F-4 were en route to Africa for use by I./JG 27, and from that point on the Emil's days in this campaign as a frontline weapon of war were numbered.

While the reduced armamant of the *Friedrich* did not endear it to all *Jagdflieger,* particularly those used to the better spread of fire from the wing cannon of the Bf 109E-7/Trop, the *experten* revelled in the concentrated fire from two machine guns and a centreline cannon. Skilled pilots like Marseille, Franzisket, Rödel and Neumann, all of whom had by this stage had thoroughly mastered the art of deflection shooting, could simply 'point the nose', fire and 'walk' the shells onto the target, usually with deadly recults. However, as the Allied forces increased in size, and improved and better protected American aircraft appeared, the

This view of dispersed III./JG 27 Bf 109E-4s in Sicily in May 1941 included the aircraft flown by the *Gruppen* adjutant, whose chevron marking is clearly visible on the nose of his Messerschmitt fighter. This machine also wears the distinctive *Gruppe* badge, which comprised the coat-of-arms of Jeseu, a town in East Prussia *(Weal)*

Also part of the III.*Gruppe Stab* was this E-4, flown by III./JG 27 Kommandeur, Haupt Max Dobislav. His chevron marking appears on the cowling, which has been removed to allow a 'black shirt' access to the propeller hub back-plate

Aircraft of I.*Gruppe* at Gambut, Libya, soon after their arrival in April 1941. The barren nature of most desert landing grounds can be readily appreciated from this panoramic view. Most ground operations, including refuelling, were carried out in the open, although tents soon sprung up as protection against the sun. Empty fuel drums, just like those scattered behind the Bf 109s, became a common marking point for airstrips across the Western Desert over the next two-and-a-half years *(Schroer)*

German pilots were soon at some disadvantage. Although most had followed the dictate of boring in as closely as possible to ensure fatal damage, this exposed the fighter to additional risk, and the Bf 109F was clearly outclassed in the armament department by early 1943. This problem was partially addressed with the arrival of the G-model in June 1942, and JG 27 was among the first *Geschwaderen* to re-equip with the sub-type. By 1943 the situation vis-a-vis Luftwaffe air superiority in the region had markedly deteriorated, and II./JG 27 soon traded up to the Bf 109G-6.

The G-6 was also rapidly issued to the other three *Gruppen* of JG 27, and by the time the unit abandoned its former hunting ground and returned to add its weight to Reich defence in late 1943, it was fully operational on Gustavs – most of the *Geschwader*'s machines were armed with a pair of additional MG 151 20 mm cannon in underwing gondolas. Although these guns weighed more than integral wing cannon would have done, they nevertheless gave the Bf 109G the firepower necessary for attacking the four-engined USAAF bombers that JG 27, and the rest of the *Jagdwaffe* on the Western Front, now faced on an almost daily basis.

A number of examples of the pressurized Bf 109G-5 were also pressed into service by II./JG 27, this specialised Gustav being similar to the G-6 in all respects, bar its high-altitude capable cockpit and MG 131 cowling machine guns, which replaced the MG 17s. The G-5 was issued to units in the autumn of 1943, but it was only produced in modest numbers and interspersed with the G-6 on production lines – there were no separate production runs. In common with other units that continued to fly the Bf 109, all *Gruppen* of JG 27 received the G-10, except III. *Gruppe*.

Operationally, the Bf 109G-14 was one of the most important of the later versions of the type, and like the G-5, its production was interspersed with the G-6. It went on to became the standard version until the end of

A scratched and stained drop tank rests on its 'dolly' awaiting attachment to the centreline shackles of 'White 6', whilst another I.*Gruppe* machine comes in to land behind it at Ain El Gazala in October 1941. This E-7/Trop belongs to one of the last *Staffel* within I./JG 27 to transition onto the Bf 109F, a fact denoted by the Emil's overall tan colour scheme, which lacks the dapple green blotches applied earlier in the campaign to allow the aircraft to blend in with the camelthorn bushes found near the coast

hostilities, and although the workhorse Daimler Benz DB 601A engine was retained, the aircraft featured the *erla haube* cockpit canopy, which greatly improved visibility when compared to the older 'greenhouse' type, with its heavy framing that had hindered pilots' all-round visibility since the Kondor Legion days. Both the original and the later taller wooden vertical tail units were fitted to the G-14.

Just as all the improvements made to the G-6 were intended to be incorporated on the G-14 to produce a 'standard' version (which was not completely achieved), the Bf 109K was planned as a series to bring together all the previous operational features, including pressurisation.

In the event, however, only the Bf 109K-4 was built in quantity, and among the recipient *Gruppen* were those of II., III. and IV./JG 27, this and older sub-types being flown by the 'Afrika' unit until the surrender. That Bf 109G-14s remained numerous in JG 27 during the final weeks of the war is reflected in the loss of three *Geschwader* pilots during March 1945. On the 1st Ogef Georg Karch of 7.*Staffel* went down at Rhode, and he was soon followed by Fw Heinz Zimmerman of 5.*Staffel* at Wulfen/Westfalen – both pilots were killed. Ten days later Uffz Franz Lamminger of 6.*Staffel* was reported missing over Wesel. All three pilots had been flying G-14s. During the war JG 27 claimed the destruction of 2700 Allied aircraft, 250 of which were Russian. The *Geschwader* lost 400 pilots killed, with a further 250 missing and 70 made prisoner.

Providing the best possible contrast with the E-7/Trop on the previous page is the much photographed 'Black 8' of 8./JG 27. The personal mount of Lt Werner Schroer, this pristine fighter actually wears four small yellow victory bars on its rudder, although they are all but invisible in this classic air-to-air shot. Schroer flew this sortie soon after arriving at Ain El Gazala in April specially for Luftwaffe photographers sent to Libya to record JG 27's combat debut. Due to the significance of this flight, the groundcrews at I./JG 27 have ensured that 'Black 8' wears as definitive a rendition of the official North Africa scheme for Bf 109s as possible

An operation repeated countless times while JG 27 remained in North Africa – ground runs enabled the *Geschwader*'s engine specialists to detect any problems with the Bf 109's generally robust DB 601 powerplant before the aircraft were pronounced fit to perform the early morning patrols

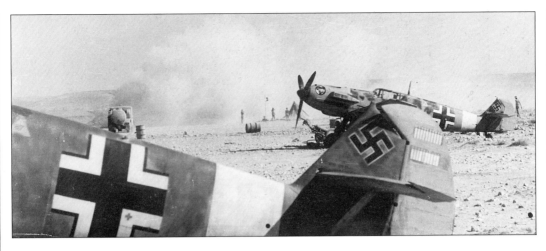

A sandstorm heralds the departure of a *schwarme* of I.*Gruppe* Bf 109E-7/Trops from Ain El Gazala in July 1941. In the foreground is 'White 1', the mount of *experte* Oblt Karl-Wolfgang Redlich, *Staffelkapitan* of 1/JG 27 – it already wore 20 kill bars on its rudder by this stage of the campaign. Although having been in action from the day the unit arrived in Libya, the Bf 109 still wears it grey European camouflage *(Weal)*

'White 1's' scheme would not have looked too dissimilar to this I./JG 27 Bf 109E-7/Trop, seen on Sicily in late March 1941 with a mechanic in its cockpit, checking revolutions and oil pressure prior to its next sortie

THE KOMMODORE'S VIEW

Oberstleutnant Eduard Neumann was Kommodore of JG 27 from April 1941 to April 1943, having led I. *Gruppe* up to June 1942. Some of his incisive and thought-provoking impressions of his unit's bloody, and ill-fated campaign in the desert follow.

'The main strength of the Bf 109E lay in its excellent performance, high diving speed and good armament. The "Emil" was superior to all enemy fighter aircraft operating in North Africa in 1941, and the initial successes gave pilots a feeling of safety and superiority, although we always felt ourselves to be inferior in numbers – quite considerably so in the latter stages of the campaign.

'The aircraft flew in the smallest possible formations (*rotte* or *schwarme*) and this offered ideal chances for gifted and aggressive pilots to show their qualities. Some, like Marseille, Homuth and others, took advantage of

this situation to a high degree. However, it is very easy to explain that, apart from the leading pilots, it was hard for newcomers and less gifted pilots to achieve similar success. The problem was often discussed in the ranks because it was clear that it was not ideal for training newcomers to become good pilots.

'On the other hand, Marseille for example, was willing to let his comrades share his tactics and rejoiced at each victory by his unit, but only a few pilots were able to gain from his experience in combat because they lacked his natural talent. This handicap was partly overcome by the morale effect on the *Geschwader* of the successes of pilots like Marseille. In fact most of the pilots in Marseille's *Staffel* acted in a support role to the "master", though this was in itself of secondary importance when compared to the results achieved.

'During the course of 1942 the *Geschwader* was equipped with the Bf 109F, and later with single examples of the Bf 109G. These types had a lot of advantages over the "Emil", but the ascendancy of performance over enemy types shrank more and more. The British received a lot of aircraft of American origin in 1942, and between June and October of that year, our own operations were very restricted for sound reasons. During these months of regeneration by the Western Desert Air Force, the *Jagdflieger* were worn out escorting the obsolete Stukas at a time when the

Almost certainly photgraphed in Sicily in late April 1941, this E-4 clearly shows how the original camouflage scheme showed through the temporary yellow paint that was applied for recognition purposes when JG 27 were first deployed to the Balkans

A Bf 109F of II.*Gruppe* shares an airfield with a Ju 88A of *Lehrgeschwader* 1. II./JG 27's Berlin bear emblem was topped by a red castle device based on the city's coat of arms *(Weal)*

By the time I. *Gruppe* received the Bf 109F, the *Geschwader*'s groundcrews were well used to open-air servicing, and they could cope supremely well with most technical problems. Here, a damaged machine is having its landing gear checked while half the servicing team hold the tail down. Judging by the missing wing-tip panels, this F-2/Trop was probably the victim of a ground loop on landing

Although not as famous as some of his fellow JG 27 *experten*, Lt Werner Schroer nonetheless flew more sorties in the Desert War than most of his peers. Amongst the first I./JG 27 pilots despatched to Libya, he literally became the backbone around which 8./JG 27 was formed. In fact, this photograph was taken during his only spell away from 'his' Staffel in three years of combat flying. Schroer was made acting I. *Geschwader*-Adjutant in November 1942, and despite only spending a brief time in the post, he still managed to use two F-4/Z Trops – this was the second of his uniquely-marked machines. Normal adjutant markings consisted of a vertical bar behind the chevron, rather than a stylised letter *(Schroer)*

Ju 87 should have been withdrawn from service. It was too slow, and therefore very difficult to cover, and such missions were very costly to JG 27. The initiative had passed to the British by mid-1942.

'The North African theatre was not considered so hard as the Channel front, but much more difficult than the Russian front. On one hand this was the consequence of the excellent fighting spirit of the British pilots and their good aircraft, and on the other, to the specific conditions of the desert. Our food was poor because the supply situation didn't function well due to the failure of the transports, and the climate affected the health of most pilots with a desert time of more than six months.

'Life on the ground was rendered more difficult by the activity of British commandos; we could expect that any night submarines would

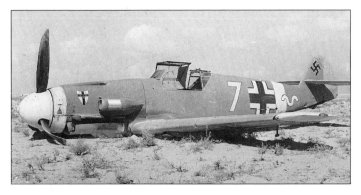

land sabotage teams, and during periods of moonlight, bombs fell. Taken together, these actions had a considerable effect.

'We really began to feel the RAF bombing attacks during the second half of 1942. It was just at the time when the German fighters were already weakened by losses and sickness; neither were the pilots very skilled in fighting bomber formations. In my opinion, this explains why the results we achieved against bombers were apparently so small.

'We saw the defensive circle used by the British fighter units until about the end of 1942; in level flight they liked to be covered by higher-flying and permanently weaving pairs. The defensive circle seemed to be a necessity because of the technical preformance of the British aircraft – they turned better than the Bf 109, but were slower in level flight and the dive. It is not too surprising that the we therefore used the advantages of our aircraft to avoid dogfights. I don't know if the "covering pairs" used by Allied formations were successful, but it is a fact that a formation of permanently weaving and turning aircraft was more visible from a greater distance than the looser, straight-flying formations used by the Germans.

'We did not find it difficult to differentiate between the Hurricane and the Spitfire, or to recognise them from American types. These latter types did cause us recognition difficulties, however, and I believe that we never did learn to distinguish between a Tomahawk, Kittyhawk or Warhawk. If these were large mistakes, the reason can only have been the uncertain

Although this photograph is regrettably undated, the spotless condition of the crestfallen III.*Gruppe* **F-4/Z Trop suggests that it was taken soon after the unit arrived in Libya in December 1941. In fact this fighter even lacks the mandatory exhaust staining over its wing roots, which probably means that it was lost on one of its first flights in North Africa. This unit was the least-known of JG 27's** *Gruppen,* **and this marvellous shot shows off its markings upon its arrival in the desert**

Marseille glances over at his *Staffel-***mates to check on the progress of their pre-flights prior to firing up his own Bf 109F-4/Z Trop**

conditions of the light, the angle of approach, or haze from the sand.

'I have spoken so far only of the German fighter pilots but not of the Italians. We were on friendly and often cordial terms with our Italian comrades, so far as was allowed by their higher staff. They were always willing to co-operate closely with us, and when we flew together during Stuka missions, for example, the Italian provided valuable support. That Italian aircraft were never good enough – they had to fly to Tobruk in 1941 in Fiat CR.42 biplanes – and that the authorities in Italy seemed not to approve of too good relations between the Germans and Italian troops at the front, is another thing entirely.'

Marseille's first Bf 109F-4/Z Trop (Werk-Nr. 12 593) is refuelled and re-armed between sorties at Martuba in February 1942

EXPERTEN

Of the hundreds of pilots who saw service in the theatre covered by this book, there is space only to mention three of those who rose to become outstanding exponents of the art of the aerial hunter. It should be borne in mind that in general, most of the *experten* were moved from one unit to another, and served on different war fronts during their long careers.

Major Hartmann Grasser of JG 51 – Beginning his combat career as a *Zerstörer* pilot with ZG 52, Grasser flew the Bf 110 with ZG 2 during the *Kanalkampf* and emerged with five victories. In February 1941 he was posted to *Stab*/JG 51 as adjutant to Werner Moelders and became Kommandeur of II. *Gruppe* for the start of the Russian campaign in June 1941. He then had seven victories to his credit. By September Grasser's score had risen to 29 and he received the *Ritterkreuz*. He continued to lead II. *Gruppe* when it moved to North Africa, and found immediate success, increasing his score to 103 by 31 August 1943, and duly receiving the coveted *Eisenlaub*. Taking up a staff appointment in Paris, Grasser later joined JG 1 in the last weeks of the conflict, and by war's end had flown 700 sorties, although he failed to increase his score. Having served on the Eastern Front, he was imprisoned in Russia for four years after the war.

Major Hartmann Grasser

Hauptmann Wolfgang Tonne of JG 53 – A member of 3./JG 53 during the Battle of France, Wolfgang Tonne remained with this unit for the assault on England. He then spent a long period of time on the Russian Front, having already achieved five victories by the time he began combat operations in the East. His tally increased until he recieved the *Ritterkreuz* on 6 September 1942 with his score at 54. In November Tonne transferred North Africa. His 100th victory came on 24 September, whereupon he was decorated with the *Eisenlaub*. Tonne was killed attempting to land at Protville, Tunisia, on 20 April 1943. He scored 122 victories and flew 641 sorties during his two-and-a-half years of uninterrupted combat flying, with all but 26 of his kills being achieved in the East.

Hauptmann Wolfgang Tonne

Leutnsnt Johann Pichler of JG 77 – Joining 7.*Staffel* in August 1940, Pichler was among the pilots who took a long time to score his first victory. This came at last on 16 May 1941 when he shot down a Hurricane. Operating the Bf 109 from rough landing grounds was a hazard that even the best pilots could not always compensate for, particularly if their machine had sustained battle damage. Johann Pichler experienced such a crash-landing on 22 May when his Bf 109E-7 sustained 85 per cent damage at Molais. As Pichler's sorties built up, JG 77 saw combat over Russia, Italy and Rumania, as well as North Africa, and his victory tally increased to an impressive 75, including 29 in the East and 16 four-engined bombers in the West. Hospitalised in Rumania, Johann Pichler fell into Russian hands on 30 August 1944. The award of the *Ritterkreuz* for his 75th victory was announced on 7 September, although due to his imprisonment Pichler never physically received his medal.

Leutnant Johann Pichler

THE MAKING OF AN *EXPERTEN*

When the French first coined the term 'ace' to distinguish a pilot who had shot down a number of enemy aircraft in aerial combat during World War I, the Germans, in common with their opponents, set a stipulation that ten kills had to have been achieved before a pilot could call himself an ace. Both sides fostered this record because it enabled daring aerial exploits – still very much a new phenomenon in the minds of the general public in 1917 – to be written about in the press of the day. Aces rose above the carnage on the ground and appeared to give warfare a new dimension that seemingly harked back to the days when knights engaged in jousts to settle their differences. With an implication that air combat was more individualistic and chivalrous than the mass slaughter indulged in by opposing armies, the idea of the ace was more or less established in aviation lore.

Neither side forgot the ace when World War 2 broke out, and the Luftwaffe continued to regard ten kills in the air as the baseline for the ace, or *experten*, and the award of medals and/or promotion. Building on this score usually brought further accolades, and in many cases the award of the coveted *Ritterkreuz* or Knight's Cross, supplemented by the *Eisenlaub* (Oak Leaves), the *Schwerin* (Swords) and *Brillianten* (Diamonds).

These awards were not, however, automatic. In an effort to prevent Germany's highest decorations from being cheapened, the stipulation for the number of fighter kills was continually raised and a system of points introduced. As the French had found in World War 1, it was easier to score victories on some fronts than on others, and the Luftwaffe made a distinction between successes over the Western and Eastern Fronts. The reason for the disparity between the German, French and British yardstick of ten kills, as opposed to the Americans who recognised five as the number in World War 1, was simply one of timing. Many Allied and German pilots had already scored more than ten victories when the term was coined in 1917, whereas the US Air Corps had just entered the war. It appeared to the American command, with the end of the conflict in sight, to be unrealistic to set the baseline score too high, so five it was. This has remained the universally recognised figure ever since.

An informal view of JG 27's greatest *experte* as he leaves 'Yellow 14' in the capable hands of the ground crew. The serious exhaust staining over the fuselage of this F-4/Z Trop suggests some vigourous throttle movement during its last sortie! As his fame spread, much of Marseille's time whilst on the ground in early 1942 was taken up answering fan mail from Germany, and entertaining Luftwaffe 'top brass'. However, come the bitter autumn he was left to fend for himself

THE APPENDICES

Luftwaffe Bf 109 fighter dispositions in the Mediterranean theatre between 1941 and 1944 – strengths at selected key periods of the campaign

		NUMBERS AVAILABLE	
		on strength	serviceable
1.	The 1941 build-up a)	14	11
	b)	36	28
	c)	163	134
	d)	154	98
2.	Pre-Alamein (20/9/42)	194	125
3.	Nearing the end in Tunisia (10/3/43)	235	137*
4.	In defence of Sicily (10/7/43)	324	182
5.	The Anzio bridgehead (3/4/44)	–	93
6.	The end in Italy (5/9/44)	54	37

*Excluding Eastern Med/Balkans

1. Bf 109 fighter strength in-theatre in detail – 1941

a) First Bf 109s in Mediterranean area were those of Müncheberg's *Staffel*, arriving on Sicily as part of *X.Fliegerkorps* for attack on Malta as of 9/2/41. First appearance on Order of Battle for 22/3/41 reads:

X.Fliegerkorps (10th Air Corps)

	on strength	serviceable
7./JG 26 Gela (Sicily)	14	11

b) First Bf 109s in North Africa belonged to I./JG 27, operational as of 14/4/41. Order of Battle for 26/4/41 reads:

X.Fliegerkorps (10th Air Corps)

	on strength	serviceable
7./JG 26 Gela (Sicily)	15	10

Fliegerführer Afrika (Air Command Africa)

	on strength	serviceable
I./JG 27 Ain El Gazala	21	18

Overall total:	36	28

c) In May, after the conclusion of the Balkans campaign, Bf 109 units of *VIII.Fliegerkorps* assembled in Greece for the invasion of Crete. Bf 109 presence in the Mediterranean area for 17/5/41 reads:

X.Fliegerkorps (10th Air Corps)

	on strength	serviceable
7./JG 26 Gela (Sicily)	15	15

Fliegerführer Afrika (Air Command Africa)

	on strength	serviceable
I./JG 27 Ain El Gazala	29	22

VIII.Fliegerkorps (8th Air Corps)

		on strength	serviceable
Stab JG 77	Molaoi (Greece)	6	5
II./JG 77	Molaoi (Greece)	3	43
III./JG 77	Molaoi (Greece)	42	33
I.(J)/LG 2	Molaoi (Greece)	28	26

Overall total:			134

d) In June 7./JG 26 joined I./JG 27 briefly in North Africa, before returning to France. *VIII.Fliegerkorps* likewise departed the Mediterranean theatre in preparation for the imminent attack on Russia. For the remainder of the year, JG 27's presence in North Africa gradually built up, with II./JG 27 arriving in September and III./JG 27 in December. JG 53 sent to Sicily to resume operations against Malta by the end of year. Thus, as of 27/12/41, Bf 109 strength stood at:

Fliegerführer Afrika (Air Command Africa)

		on strength	serviceable
Stab JG 27	Arco	3	3
I./JG 27	Arco	24	10
II./JG 27	Arco	22	20
III./JG 27	Arco	20	10

II.Fliegerkorps (2nd Air Corps) – replaced *X.Fliegerkorps* on Sicily

		on strength	serviceable
Stab JG 53	Comiso (Sicily)	6	6
I./JG 53	Gela (Sicily)	40	31
II./JG 53	Comiso (Sicily)	39	28
III./JG 53	Catania (Sicily)	0	0*

(*Bf 109s passed to JG 27 when *Gruppe* retired from service in Libya and returned to Sicily to re-equip)

Overall total:		154	98

2. Bf 109 presence in the Mediterranean Theatre as of 20/9/42 (at the height of the Western Desert campaign, one month prior to El Alamein)

Luftflotte 2 (Air Fleet 2)

Fliegerführer Afrika (Air Command Africa)

		on strength	serviceable
Stab JG 27	Sanyet/Quotaifiya	3	2
I./JG 27	Turbiya	28	15
II./JG 27	Sanyet	26	16
III./JG 27	Sanyet/Quasaba	28	18
III./JG 53	Quasaba East	27	14
Total:		**112**	**65**

II.Fliegerkorps (2nd Air Corps) on Sicily

Stab JG 53	Comiso	5	5
II./JG 53	Comiso	32	25
I./JG 77	Comiso	34	24
Total:		**71**	**54**

X.Fliegerkorps (10th Air Corps) in Greece and on Crete

Jagdkommando 27*	11	6
Kastelli (Crete)		
(Fighter Detachment 27)		

(*Also known as '*Jagdkommando Kreta*')

Overall Total:	**194**	**125**

3. Bf 109 presence in the Mediterranean as of 10/3/43*
(*towards the end in Tunisia and final evacuation of Africa)

Luftflotte 2 (Air Fleet 2)

II.Fliegerkorps (2nd Air Corps)

		on strength	serviceable
II./JG 27	Sicily	31	18
9./JG 53	Sicily	8	5

Fliegerführer Afrika (Air Command Africa)

Stab JG 53	Tunisia	3	3
I./JG 53	Tunisia	36	19
II./JG 53	Tunisia	33	25
III./JG 53*	Tunisia	47	26
(*minus 9.*Staffel*)			
Stab JG 77	Tunisia	3	2
I./JG 77	Tunisia	26	17
II./JG 77	Tunisia	24	11

III./JG 77	Tunisia	24	11
Overall total:		**235**	**137***

(*excluding *X.Fliegerkorps* in eastern Mediterranean

4. Bf 109 presence in the Mediterranean as of 10/7/43 (preparing for the defence of Sicily)

Luftflotte 2 (Air Fleet 2)

(all units deployed on Sicily or in southern Italy)

	on strength	serviceable
IV./JG 3	36	28
II./JG 27	22	14
Stab JG 53	6	2
I./JG 53	36	15
II./JG 53	22	18
III./JG 53	30	12
Stab JG 77	3	2
I./JG 77	39	18
II./JG 77	35	3
III./JG 77	36	30

Lw.Kdo.Südost (Air Force Command South-East)

X.Fliegerkorps (10th Air Corps)
(all units deployed Greece/Crete)

Stab JG 27	2	1
III./JG 27	28	19
IV./JG 27	29	20
Overall Total:	**324**	**182**

5. Bf 109 presence in the Mediterranean as of 3/4/44 (at the time of the Anzio bridgehead battle)

Luftflotte 2 (Air Fleet 2)

Fliegerführer Afrika (Air Command Africa)
Jagdabschnittsführer Süd (Fighter Sector Leader South)

		serviceable
III./JG 53	Arlena	15
II./JG 77	Drago	15

Jagdfliegerführer Oberitalien (Fighter Leader Upper Italy)
Jagdabschnittsführer Ost (Fighter Sector Leader East)

Stab JG 53	Tricesimo	2
I./JG 53	Maniago	20
Stab JG 77	Idome	2
I./JG 77	Lavariano	27

Jagdfliegerführer Mitte (Fighter Sector Leader Central)

I./JG 4*   Ferrara   0
(*re-equipping)

Lw.Kdo.Südost (Air Force Command South-East)

Kom.Gen.d.dtsch.Lw.i.Griechenland (Luftwaffe C.-in-C Greece)

7./JG 27
1 Schwarme: Kalamaki (Greece)   4
1 Schwarme: Malames (Crete)   4
1 Schwarme: Gaddura (Rhodes)   4

Overall Total:	**93**

6. Bf 109 Fighter Strength in the Mediterranean in 1944 (towards the close of the Italian Campaign)

Final appearance of Luftwaffe Bf 109s on Mediterranean Order of Battle dated 5/9/44

Luftflotte 2 (Air Fleet 2)

	on strength	serviceable
II./JG 77	54	37
Ghedi (northern Italy)		

Thereafter, only Bf 109s in Mediterranean (northern Italy) were reconnaissance machines of the Luftwaffe and Italian Republic Gruppi*)
(*as of 9/4/45)

I° Gruppo Caccia:	45	37
II° Gruppo Caccia:	32	16
III° Gruppo Caccia:	21	13 (non-op)
Overall Total:	**98**	**66**

JAGDFLIEGER OF LUFTWAFFE FIGHTER UNITS IN THE NORTH AFRICAN & MEDITERRANEAN THEATRES

oa – score on arrival in theatre
fs – final score in desert
RK – Ritterkreuz (Knight's Cross) holders
E – Eisenlaub (Oak Leaves)

Name & rank	score	notes

7./JG 26

(Red heart marking) – Sicily (Gela) 9 Feb to late Aug 1941; apparently 12 pilots arrived, and the following nine are known:

Name & rank	score	notes
Ehlen, Uffz Karl Heinz		
Haugk, Helmut	6	(from III.Gr)
Johannsen, Fw Hans		
Kestel, Fw Melchior		
Kuhdorf, Fw Karl		
Laub, Gefr Karl		
Liebing, Uffz ?		
Mietusch, Lt Klaus		
Müncheberg, Oblt Joachim	23 oa/135 fs	RK (KIA 23/3/43)

I/JG 27

(Bf 109E) to Libya 18 Apr 41, to Germany 12 Nov 42 (with Stab/JG 27); received Bf 109F-4 Sept 41

Name & rank	score	notes
Espenlaub, Obfw Albert	14	PoW 13 Dec 41
Elles, Fw Franz	5 fs	
Forster, Oberfw Hermann	6 oa	
Franzisket, Oblt Ludwig (Adj)	14 oa/37 fs	RK
Geisshardt, Fritz	10 fs	
Grimm, Joseph	5 fs	
Hoffmann, Freidrich	11 fs	
Homuth, Oblt Gerhard	15 oa/46 fs	RK (3.St Kap; Kdr)

Name & rank	score	notes
Kaiser, Emil	5 fs	
Keppler, Gerhard	6 fs	
Kothmann, Lt Willi	7 oa	
Körner, Friedrich	36 fs	
Kowalski, Herbert	5 fs	
Maak, Ernst	6 fs	
Marseille, Oberfh Hans-Joachim	7oa/158 fs	RK/E KIA 30 Sept 42 (as Hptmn)
Mentnich, Karl	5 fs	
Neumann, Hptm Eduard	9 oa	(Kdr/Kdre May 42)
Remmer, Hans	14 fs	
Redlich, Oblt Karl-Heinz	10 oa	RK (1.St Kap) KIA 20/5/45
Schmidt, Heinz	5 fs	
Schneider, Hugo	9 fs	
Schroer, Lt Werner	61 fs	
Sinner, Rudolf	32 fs	(also II.Gr)
Stahlschmidt, Lt Hans-Arnold	59 fs	11.St Kap KIA 7 Sept 42
Steinhausen, Günther	40 fs	KIA 6 Sept 42
von Lieres u Wilkau, Karl	24 fs	

II./JG 27

(with Bf 109F Sept 41; Bf 109G – 6 Nov 42; to Italy late Nov 42)

Name & rank	score	notes
Bendert, Karl-Heinz	36 fs	
Borngen, Ernst	13 fs	
Clade, Emil	9 fs	
Düllberg, Ernst	10 fs	
Gerlitz, Hptm Erich (Kdr)	12 fs	to JG 53 5/42
Heidel, Alfred	7 fs	
Jenisch, Kurt	9 fs	
Kientsch, Willi	15 fs	
Krenzke, Herbert	6 fs	
Lippert, Hptm Wolfgang	25 oa/29 fs	(Kdre) PoW 23 Nov 41
Niederhofer, Hans	13	
Reuter, Uffz Horst	21	PoW May 42
Rödel, Oblt Gustav	20 oa/52 fs	(4.St Kap) RK
Sawallisch, Obfw Erwin	19 oa	RK
Schneider, Lt Bernd	14 fs	KIA 29 Apr 43
Schulz, Obfw Otto	9 oa/51 fs	posted 2/42; RK; KIA 17 June 42
Steis, Heinrich	12 fs	
Stiegler, Franz	17 fs	
Vögl, Ferdinand	25 fs	

III./JG 27

(5 May – end May 1941) Sicily; to N Africa Dec 41; to Crete/Greece 12 Nov 42)

Name & rank	score	notes
Kageneck, Oblt Erbo Graf von	67 fs	RK/E, KIA 23 Dec 41

Stab./JG 27

(Dec 41; to Crete/ Greece 12 Nov 42)

Name & rank	score	notes
Woldenga, Maj Bernhard		(Kdre JG 27)

II./JG 3

(From Dec 41; to USSR May 42)

Name & rank	score	notes
Krahl, Hptm Karl-Heinz	20+ oa	RK KIA (Malta) 14 Apr 42
Ohlrogge, Fw Walter	39 oa	RK
Kirschner, Lt Joachim		
Brändle, Oblt Kurt	40+ oa	(origin 4./JG 53 St Kap, KIA 24 Apr 43
Schwager, Uffz Franz		

I./JG 53

(to Sicily late Nov 41; to USSR May 42; to Sicily Oct 42)

Name & rank	score	notes
von Maltzahn, Maj Günther	50+ oa	RK/E
Schiess, Lt Franz	15 oa	(8.St Kap), KIA 2 Sept 43
Kaminski, Hptm Herbert		(Kdr)
Quaet-Faslem, Lt Klaus	5+ oa	(Adj)
Müller, Oblt Freidrich-Karl 'Tutti'	20 oa	(1.St Kap) Kdre as Hptm Nov 42 with 100 +
Tonne, Lt Wolfgang	101oa/122fs	KIA 20 Apr 43
Crinius, Lt Wilhelm	100oa/114 fs	RK/E PoW 13 Jan 43
Möller, Lt Hans	30+	(St Kap) PoW 25 Mar 43

II./JG 53

(to Sicily late Nov 41)

Name & rank	score	notes
Michalski, Lt Gerhard	20	(4.St Kap)
Dinger, Lt Fritz	67 fs	RK, KIA 23 Nov 42
Rollwage, Fw Herbert		RK

III./JG 53

(to Sicily late Nov 41; to Libya Dec 41; to Sicily Dec 41; to Libya 20 May 42; to Sicily Oct 42)

Name & rank	score	notes
Altendorf, Oblt Heinz	14	(7.St Kap) PoW 15 Dec 41
Bahnsen	6/7+	
Harder, Jürgen	16 fs	

Name & rank	score	notes
Götz, Oblt Franz	30 oa+	(8.StKap)
Klager, Ernst	5+	
Neuhoff, Lt Hermann	37 oa/40 fs	RK, PoW 10 Apr 42
Schramm, Lt Herbert	37 oa	
Seidel,	5+	
Stumpf, Obfw Werner	15 oa/47 fs	RK, KIA Oct 42
Wilcke, Hptm Wolf-Dietrich	33 oa	RK, (to JG 3)

I./JG 77

(to Sicily 6 July 42; to N Africa)

Name & rank	score	notes
Bär, Hptm Heinz	120 oa/130 fs	RK/E/Sch
Berres, Heinz-Edgar	7 oa/53 fs	(1.St Kap) KIA 24/7/43
Freytag, Oblt Seigfried	49+ oa	RK
Geisshardt, Hptm Fritz	82 oa	RK/E

II./JG 77

(to N Africa Dec 42)

Name & rank	score	notes
Badum, Lt Johannes	54 fs	KIA 12 Jan 43
Burckhardt, Oblt Lutz-Wilhelm	53 oa	
Hackl, Hptm Anton	118 oa	
Mader, Hptm Anton	50+ oa	
Reinert, Lt Ernst-Wilhelm	104 oa	

III./JG 77

(to N Africa from USSR 28 Oct 42)

Name & rank	score	notes
Goedert, Oblt Helmut	25 oa	
Huy, Hptm Wolf-Dietrich	40 oa	PoW 29 Oct 42
Kaiser, Herbert	7	
Omert, Oblt Emil	55 oa	
Ubben, Hptm Kurt	95 oa; 101	(Kdre)

Stab./JG 77

(to N Africa from USSR 28 Oct 42)

Name & rank	score	notes
Müncheberg, Joachim (Kdre)	100+ oa/135 fs	KIA 23 Mar 43
Steinhoff, Maj Johannes	150 oa	RK/E

II./JG 51

(to N Africa 14 Nov 42; to Sicily 20 Apr 43)

Name & rank	score	notes
Grasser, Hptm Hartmann	92 oa	
Hafner, Fw Anton	62 oa/82 fs	KIA 17/10/44
Heydrich, Oblt Hans		KIA 12/1/43

Name & rank	score	notes
Mink, Obfw Wilhelm		KIA 12/3/45
Rammelt, Maj Karl	5+/46 fs	(Kdre)
Puschmann, Hptm Herbert	5+	

II./JG 2

(from France to N Africa Nov 42) Fw 190A; (to France Mar 43)

Name & rank	score	notes
Bühligen, Oblt Kurt	40	(N Africa)
Rudorffer, Lt Erich	27	(N Africa)
Dickfeld, Oblt Adolf	18	(N Africa)
Goltzsch, Obfw Kurt	14	(N Africa)

11./JG 2

(to Sicily Nov 42); Bf 109 (reinforcement for II./JG 53)
Meimberg, Oblt Julius

11./JG 26

(to N Africa Nov 42; reinforcement for II./JG 51)

Name & rank	score	notes
Westphal, Oblt Hans-Jürgen	22 fs	(11.St Kap)

SPECIFICATIONS

Messerschmitt Bf 109E-7/Trop
Type: single-seat fighter
Armament: Two MG FF 20 mm cannon and two MG 17 7.9 mm machines guns
Powerplant: one Daimler-Benz DB 601A inverted-vee engine rated at 1050 hp for take-off
Dimensions: span 32ft 4$\frac{1}{2}$ in; length 28 ft 4$\frac{1}{2}$ in; height 11ft 2$\frac{1}{3}$ in
Weights: maximum loaded 5523 lb; empty 4421 lb;
Performance: maximum speed 354 mph; cruising speed 298 mph; service ceiling 36,091 ft; range 412 miles

Messerschmitt Bf 109F-4
Type: single-seat fighter
Armament: One MG 151 20 mm cannon and two MG 17 7.9-mm machine guns
Powerplant: one Daimler-Benz DB 601E inverted-vee engine rated at 1350 hp for take-off
Dimensions: span 32 ft 5$\frac{3}{4}$ in; length 29 ft; height 8 ft 6 in
Weights: maximum loaded 6393 lb; empty 5269 lb
Performance: maximum speed 334 mph; cruising speed 310 mph; service ceiling 39,370 ft; range 442 miles

Messerschmitt Bf 109G-6
Type: single-seat fighter
Armament: One Mk 108 30 mm or MG 151 20 mm cannon and two MG 17 7.9 mm machine guns, plus optional fit of 2 x 20 mm cannon in underwing gondolas
Powerplant: one Daimler-Benz DB 605AM inverted-vee engine rated at 1475 hp for take-off
Dimensions: span 32 ft 6$\frac{1}{2}$ in; length 29 ft; height 8 ft 2$\frac{1}{2}$ in
Weights: maximum loaded 7491 lb; empty 5893 lb
Performance: maximum speed 340 mph; cruising speed 310 mph; service ceiling 37,890 ft; range 350 miles

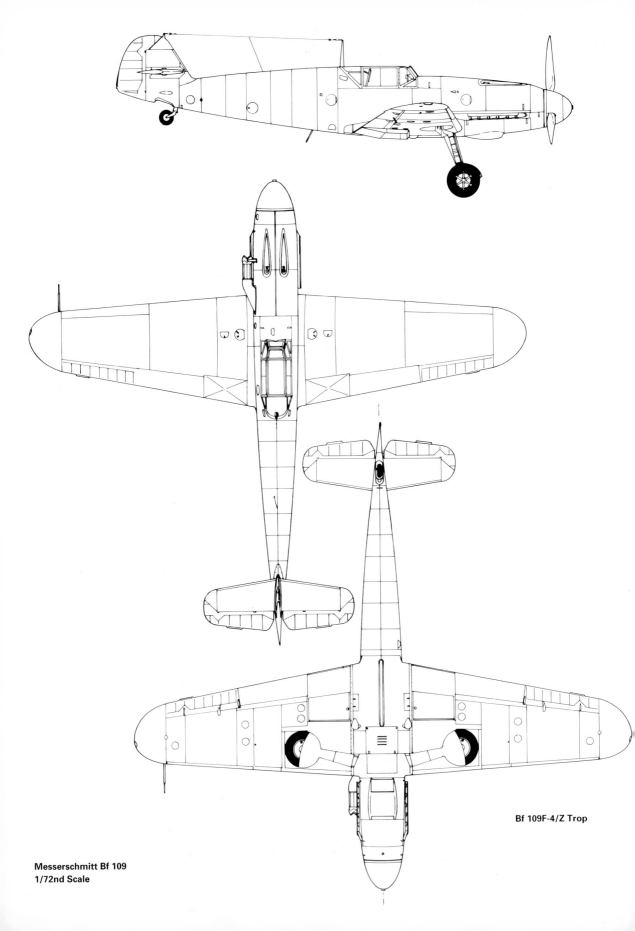

Bf 109F-4/Z Trop

Messerschmitt Bf 109
1/72nd Scale

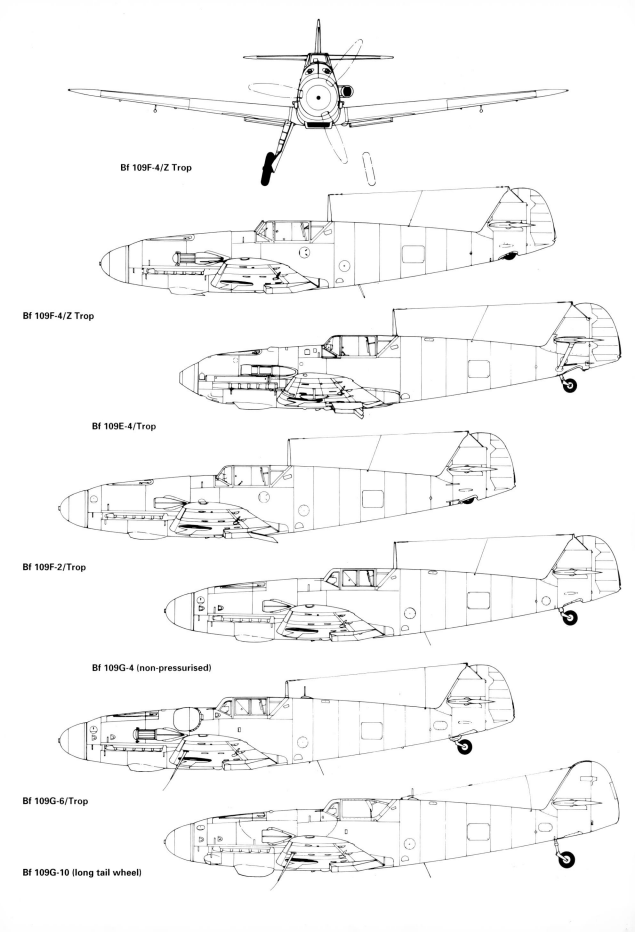

Bf 109F-4/Z Trop

Bf 109F-4/Z Trop

Bf 109E-4/Trop

Bf 109F-2/Trop

Bf 109G-4 (non-pressurised)

Bf 109G-6/Trop

Bf 109G-10 (long tail wheel)

COLOUR PLATES

1
Bf 109G-1 'White 11', flown by Oberleutant Julius Meimberg, *Staffelkapitan* 11./JG 2, Bizerta/Tunisia, November 1942

Activated in France in June 1942 as a specialized high-altitude *Staffel* flying the pressurized Bf 109G-1, 11./JG2 transferred to the Mediterranean in November. Initially retaining their standard European three-tone grey (74/75/76) camouflage, as shown here on Werk-Nr.14 063, the *Staffel* was incorporated into II./JG 53 almost immediately upon its arrival in Tunisia. In April 1944 'Jule' Meimberg himself assumed command of the *Gruppe*, which he then led until the war's end, gaining a final total of 53 kills.

2
Bf 109F-4/Z Trop 'White Chevron/Triangle', flown by Hauptmann Karl-Heinz Krahl, *Gruppenkommandeur* II./JG 3, San Pietro/Italy, April 1942

Like JG 2 'Richthofen' above, JG 3 'Udet' was also called upon to provide fighter reinforcement for the Mediterranean. In the latter's case, however, it was an entire *Gruppe*, II./JG 3, which was sent south early in 1942. Their aircraft received the standard Mediterranean camouflage scheme of Sandbraun (Tan) 79 over Hellblau (Light Blue) 78; some, like Werk-Nr.8665 here, also being given a light overspray of Green 80. Krahl was killed over Malta on 14 April 1942, brought down by AA fire during a low-level sweep, with his score standing at an estimated 24 victories.

3
Bf 109F-4/Z Trop 'Yellow 3', flown by Unteroffizier Franz Schwaiger, 6./JG 3, Castel Benito/Libya, February 1942

Another II./JG 3 machine in standard theatre finish, complete with black and white gyronny *Gruppe* badge, outlined in red. Personal markings, such as the name of the pilot's girlfriend on the cowling of this fighter, were relatively uncommon at this stage in the Mediterranean. Schwaiger's score had risen to 67 when, as *Staffelkapitan* of 1./JG 3, he lost his life during the Defence of the Reich after being strafed on the ground by a P-51 following a forced-landing north of Augsburg due to lack of fuel on 24 April 1944.

4
Bf 109G-6/Trop 'Black Double Chevron', flown by Major Franz Beyer, *Gruppenkommandeur* IV./JG 3, San Severo/Italy, August 1943

After II./JG 3's brief sojourn in early 1942, the 'Udet' *Geschwader* did not reappear in the Mediterranean until the newly-activated IV.*Gruppe* was deployed to Italy from July to September 1943. Beyer's aircraft displays standard mid-war finish and *Gruppenkommandeur* chevrons, plus the winged 'U' (for Udet) *Geschwader* badge and the attenuated wavy bar on the aft fuselage, adopted by JG 3 to indicate its fourth *Gruppe*. Note also the underwing launch tubes for 210 mm air-to-air rockets. Beyer was killed on 11 February 1944 near Venlo, in Holland, when his fighter struck a tree while attempting to escape at low-level from a pair of pursuing Spitfires. His score then stood at 81 kills, plus at least 10 aircraft destroyed on the ground.

5
Bf 109E-7 'White 15', flown by Feldwebel Karl Laub, 7./JG 26, Ain El Gazala/Libya, June 1941

Another *Jagdgeschwader* which saw limited service in the Mediterranean was JG 26 *Schlageter*, whose 7.*Staffel* was deployed into the area from February to August/September 1941. Laub's aircraft is typical of the *Staffel* during that period, combining an aft fuselage white theatre band with yellow cowling and rudder, and displaying both the red heart *Staffel* badge and the black script 'S' on a white shield of the parent *Geschwader*. 'Karlchen' ('Little Charlie') Laub was one of JG 26's backbone NCO pilots, scoring seven kills before himself falling victim to RAF Tempests on 14 December 1944 while providing aerial cover for the Me 262 fighter base at Plantluenne.

6
Bf 109E-7 'White 12', flown by Oberleutnant Joachim Müncheberg, *Staffelkapitan* 7./JG 26, Gela/Sicily, February 1941

Basically similar to the machine above, 'White 12' eschewed the white theatre band in favour of retaining the vertical III.*Gruppe* bar. Note the white metal pennant attached to the aerial mast indicating the Staffelkapitan's aircraft. Müncheberg was one of the Luftwaffe's leading aces, or *experten*, he alone being responsible for 25 of the *Staffel*'s 52 Mediterranean victories during this first deployment. In October 1942 he was to return to North Africa as Kommodore of JG 77, only to lose his life on 23 March 1943 when his 135th and final victim, a US Spitfire, exploded directly in front of his Bf 109G after a particularly close burst of cannon fire. Both fighters crashed in flames

7
Bf 109G-1 'Black 1', flown by Oberleutnant Hans-Jürgen Westphal, *Staffelkapitan* 11./JG 26, Trapani/Sicily, November 1942

11./JG 26 was the second specialized high-altitude *Staffel* activated in France (in August 1942), and subsequently deployed to the Mediterranean the following November. Westphal's machine depicts all the characteristics of a standard pressurized G-1. Note, however, the even more truncated wavy bar aft of the fuselage cross, which was 11./JG 26's *Staffel* marking. The unit suffered heavily in a bombing raid on their Tunis base on 3 December 1942, after which most of the few surviving pilots, and their aircraft, were incorporated into II./JG 51. Westphal, himself, survived the war with 22 victories.

8
Bf 109F-4/Z Trop 'White Chevron A-bar', flown by Hauptmann Werner Schroer, Acting Adjutant JG 27, Martuba/Libya, circa November 1942

JG 27 is the *Geschwader* synonymous with Luftwaffe fighters in the Western Desert, and Schroer's machine provides a perfect example of desert finish and markings; faded tan (79) upper surfaces and light blue (78) undersides, combined with a full set of white theatre markings – nose and spinner, aft fuselage band, and wingtips. As acting *Geschwader*-Adjutant, Schroer has deviated from the regulation Stab symbol, however, by replacing the usual vertical bar behind the adjutant's single chevron with a stylized 'A'.

9
Bf 109F-4/Z Trop 'Black Chevron T', flown by Oberleutnant Rudolf Sinner, Technical Officer JG 27, Martuba/Libya, circa April 1942

Although an obvious stablemate to the machine above, the demarcation line between the upper and lower camouflage colours on Sinner's F-4 comes much further down the fuselage sides. Note also the different, shallower, style of the (black) chevron, plus another deviation from the norm by the substitution of a 'T' within the arms of the chevron in place of the TO's regulation symbol of chevron/vertical bar/circle. Also of interest are the three black victory bars on the rudder, the first of 39 kills to be claimed by war's end.

10
Bf 109F-2/Trop 'Black Chevron/Triangle', flown by Hauptmann Eduard Neumann, *Gruppenkommandeur* I./JG 27, Martuba/Libya, circa December 1941

Another standard finish and set of markings, combined this time with yellow nose and rudder. Note I.*Gruppe*'s well-known badge. Although now firmly associated with the Western Desert, this motif had infact already being worn by its Bf 109Es in France well before JG 27 was ever

contemplated for service in the Mediterranean. Its introduction in 1940 was not an early example of ESP, just one of coincidence; it merely reflected the first *Gruppenkommandeur*'s interest in Germany's one-time African colonies! 'Edu' Neumann, himself, subsequently rose to command JG 27, and survived the war with 13 victories to his credit.

11
Bf 109E-7/Trop 'Black Chevron A', flown by Oberleutnant Ludwig Franzisket, *Gruppen-Adjutant* I./JG 27, Castel Benito/Libya, April 1941

Franzisket's first E-7/Trop is representative of JG 27's early aircraft in the Mediterranean; still bearing European two-tone green (70/71) upper surfaces and light blue (65) undersides, but with the aft fuselage white theatre band already applied. Note the light overspray of green dapple on the yellow cowling and rudder, and the use of a small capital 'A' within the chevron to indicate Franzisket's role of adjutant. The 14 white victory bars on the rudder would rise to 43 by the cessation of hostilities.

12
Bf 109E-7/Trop 'Black Chevron', flown by Oberleutnant Ludwig Franzisket, *Gruppen-Adjutant* I./JG 27, Ain El Gazala/Libya, circa October 1941

A later E-7/Trop flown by Franzisket provides one of the few examples of an E-series (other than the *Jabos* of SKG 210) to be finished in the tan (79) scheme (seen here in its darker, unfaded state). Note too that the 'A' has now been dropped, leaving just the regulation single chevron *Gruppen*-Adjutant symbol. Promoted to Major, Franzisket commanded JG 27 during the closing months of the war.

13
Bf 109F-2/Trop 'White 11', flown by Oberfeldwebel Albert Espenlaub, 1./JG 27, Martuba/Libya, December 1941

Displaying the 14 victory bars of the kills achieved since his arrival in the desert, Espenlaub's standard-finish F-2/Trop crash-landed in Allied territory on 13 December 1941 following a fierce dogfight with RAF Hurricanes, and its pilot was quickly taken prisoner by 8th Army troops literally minutes after his stricken Messershcmitt had come to a grinding halt. Not overly keen about having to remain behind barbed wire for the rest of the war, Espenlaub attempted to give his captors the slip at the first available opportunity, and only hours after becoming a PoW, was unceremoniously shot dead whilst making his bid for freedom later that very same day.

14
Bf 109E-7/Trop 'White 1', flown by Oberleutnant

Karl-Wolfgang Redlich, *Staffelkapitan* 1./JG 27, Ain El Gazala/Libya, July 1941

Another early E-7/Trop combining Euroepan camouflage with a white Mediterranean theatre band, Redlich's aircraft displays 20 victory bars on its yellow rudder. Having scored his first 4 kills with the Legion Condor in Spain, his 43rd, and last, victim was a B-24 destroyed over Austria on 20 May 1944 in an engagement which Redlich, by then serving as *Gruppenkommandeur* of I./JG 27, did not himself survive.

15

Bf 109F-4/Z Trop 'Red 1', flown by Leutnant Hans-Arnold Stahlschmidt, *Staffelkapitan* 2./JG 27, Quotaifiya/Egypt, August 1942

'Fiffi' Stahlschmidt, a close friend of Marseille, was one of the most successful Luftwaffe fighter pilots in the Western Desert. The 48 victories depicted here had already won him the Knight's Cross. On 7 September 1942, with another 11 kills to his credit, he was reported missing in action after being bounced by Spitfires south-east of El Alamein.

16

Bf 109F-2/Trop 'Yellow 1', flown by Oberleutnant Gerhard Homuth, *Staffelkapitan* 3./JG 27, Martuba/Libya, February 1942

Almost a replica of Stahlschmidt's F-4, above, in terms of markings, except for the absence of a white nose band and the different *Staffel* colour of the individual numeral. The scoreboard on Homuth's rudder bears witness to his being another of the desert's leading *experten*. He was to survive the African campaign, and add another 23 victories to the 40 seen here, before being shot down over the Russian Front on 3 August 1943 while serving as *Gruppenkommandeur* of I./JG 54.

17

Bf 109F-4/Z Trop 'Yellow 14', Leutnant Hans-Joachim Marseille, 3./JG 27, Martuba/Libya, February 1942

Just as JG 27 is one *Jagdgeschwader* indelibly linked with Luftwaffe fighter operations in North Africa, so the name of one member of that *Jagdgeschwader* must stand out above all others. But Hans-Joachim Marseille's arrival in the desert was not an auspicious one. As one of his superior officers has since publicly stated, 'He brought a very bad military reputation along with him, and was prone to showing off.' But the rigours and hardships of North African campaigning wrought an amazing change in the long-haired youth from Berlin, who liked to move in film-star circles; a change that set him on a meteoric path to become the greatest ace of the desert war. The 50 victory bars on the primer red rudder of Werk-Nr.12 593, the last of which earned him the Knight's Cross on 22 February 1942, were just the beginning.

18

Bf 109F-4/Z Trop 'Yellow 14', flown by Leutnant Hans-Joachim Marseille, 3./JG 27, Tmimi/Libya, May 1942

Three months later, Werk-Nr.10 059 saw Marseille's personal score climb to 68.

19

Bf 109F-4/Z Trop 'Yellow 14', flown by Oberleutnant Hans-Joachim Marseille, *Staffelkapitan* 3./JG 27, Ain El Gazala/Libya, June 1942

In just over a month those 68 had rocketed to 101. By now promoted to *Staffelkapitan*, Marseille opted to retain the lucky 'Yellow 14' on Werk-Nr.10 137, and with rudder space to accommodate the ever-growing tally of victory bars rapidly running out, the first 70 – which had gained him the Oak Leaves to his Knight's Cross – were now depicted as shown here; collectively and suitably wreathed.

20

Bf 109F-4/Trop: 'Yellow 14', flown by Hauptmann Hans-Joachim Marseille, *Staffelkapitan* 3./JG 27, Quotaifiya/Egypt, September 1942

It took just three more months to amass the next 50 kills, by which time the Swords and Diamonds had been added to his decorations. Werk-Nr.8673, the first of Marseille's F-series known to have carried the I.*Gruppe* badge, showed signs of previous ownership beneath the by-now famous 'Yellow 14'. Then, on 30 September 1942, the unthinkable happened. Returning from a routine sweep in new Bf 109G-2 Werk-Nr.14 256, the cockpit began to fill with smoke. Half-asphyxiated, Marseille attempted to bail out, only to be slammed against the aircraft's tailplane as soon as he left the cockpit and then plummet to earth with his parachute barely streaming. The greatest desert fighter of them all scored 158 victories – all but 7 of them in the skies of Africa.

21

Bf 109F-4/Trop 'Black Double Chevron', flown by Hauptmann Wolfgang Lippert, *Gruppenkommandeur* II./JG 27, Ain El Gazala/Libya, November 1941

Lippert was another JG 27 ace who had cut his teeth in Spain, where he had claimed five kills. His Werk-Nr.8469, shown here, sports standard finish, plus II.*Gruppe*'s badge (the black bear of the city of Berlin, their adopted home town), the regulation double chevron of a *Gruppenkommandeur*, and 25 victory bars on the rudder. Lippert was to score his 29th, and last, kill on 23 November 1941, being shot down later that same day. He broke both legs on impact with the tailplane while baling out over Allied lines. Despite hospitalization, gangrene set in and he died of an embolism, following the amputation of both legs, ten days after capture.

22

Bf 109G-4/Trop 'White Triple Chevron 4', flown by Hauptmann Gustav Rödel, *Gruppenkommandeur* II./JG 27, Trapani/Sicily, April 1943

Another ex-Legion Condor pilot, Oberleutnant Rödel served briefly as acting *Gruppenkommandeur* of II./JG 27 after the loss of Lippert. He subsequently assumed full command for some 11 months during 1942/43. Towards the end of this period his G-4 displayed this unusual set of Stab markings. After commanding II.*Gruppe*, Rödel became the penultimate Kommodore of JG 27, and survived the war with 98 kills.

23

Bf-109F-4/Z Trop 'Black Chevron', flown by Oberleutnant Ernst Düllberg, *Gruppen*-Adjutant II./JG 27, Tmimi/Libya, May 1942

As Adjutant of II./JG 27, Düllberg's F-4 displays a textbook camouflage finish and regulation set of markings. The only touch of individualism here is provided by the 14 kill marks on the rudder. In every sense a survivor, Dülberg ended the war as Kommodore of JG 76 in Austria, his total score having risen to exactly 50 by this time – 10 of these were four-engined bombers.

24

Bf 109F-4/Z Trop 'White 12', flown by Oberfeldwebel Franz Stiegler, 4./JG 27, Quotaifiya/Egypt, August 1942

An almost anonymous F-4, Steigler's machine carries neither the II.*Gruppe* badge on the nose or 4.*Staffel*'s own emblem of a broken-winged, crash-landing, RAF lion on its usual place below the cockpit. The undemonstrative Stiegler nonetheless achieved a total fo 28 kills, including 5 heavy bombers, by war's end.

25

Bf 109F-4/Z Trop 'Yellow 2', flown by Oberfeldwebel Otto Schulz, 6./JG 27, Tmimi/Libya, May 1942

Another standard-finish II.*Gruppe* fighter, Schulz's F-4 is personalized by its rudder scoreboard – 6 kills prior to the Russian campaign, 3 on the Eastern Front and 35 over Africa. He would down another 7 over the Western Desert (bringing his total to 51), before he himself fell victim to a Kittyhawk near Sisi Rezegh on 17 June 1942.

26

Bf 109F-4/Z Trop 'Yellow 1', flown by Oberleutnant Rudolf Sinner, *Staffelkapitan* 6./JG 27, Tmimi/Libya, June 1942

By contrast, 'Rudi' Sinner's score had now reached 6 (compare with profile 9), as witness the victory bars shown here on his rudder. The latter are obviously a replacement item; note the higher line of demarcation between upper and lower camouflage tones. Major Sinner survived the war, despite being wounded during the closing weeks when commanding III./JG 7, a *Gruppe* of Me 262 fighters. He finished with a score of 39 kills, all but 7 of which had been gained in the desert.

27

Bf 109E-4 'Black Chevron/Triangle', flown by Hauptmann Max Dobislav, *Gruppenkom-mandeur* III./JG 27, Sicily, May 1941

After the campaign in the Balkans, III./JG 27 were based briefly on Sicily for the attack against Malta – hence the hybrid theatre markings of a yellow cowling (a legacy from Yugoslavia and Greece), coupled with the white rudder and aft fuselage band of the Mediterranean. The *Gruppe*'s practice of placing Stab symbols and numerals on the engine cowling dates back to its previous existence as the original I./JG 1; as too does the *Gruppe* badge, comprising the coat-of-arms of Jesau in East Prussia – its home base – with three Bf 109 silhouettes superimposed. Dobislav, who had been with the *Gruppe* since its pre-war days, somehow survived the hostilities with a final total of 15 kills.

28

Bf 109G-6/R6 Trop 'Black Double Chevron', flown by Hauptmann Ernst Düllberg, *Gruppenkommandeur* III./JG 27, Argos/Greece, circa October 1943

Düllberg was another long-standing member of the *Gruppe*, whose service – broken only by a spell of duty with II.*Gruppe* (see profile 23) – dated back to the days of I./JG 1. But by the time he assumed command late in 1942, III./JG 27's idiosyncratic practice of locating individual identity markings on the cowlings of their machines had been dropped. Here, his G-6 'Kanonenboot' ('Gunboat') displays markings fully in compliance with current regulations, including all-white vertical tail surfaces to indicate a formation leader.

29

Bf 109G-6/R6 Trop 'White 9', flown by Oberleutnant Emil Clade, *Staffelkapitan* 7./JG 27, Kalamaki/Greece, January 1944

The 'Kanonenboot' of Emil Clade illustrates the 7.*Staffel* badge of an apple pierced by an arrow, the whole in a Revi gunsight (could one of Clade's immediate predecessors have had Swiss connections, perhaps?). It also marks the end of JG 27's three years of service in the Mediterranean area, for within a matter of weeks III.*Gruppe* would be retiring northwards through the Balkans to Vienna, where their white theatre bands would give way to the pale green of a Reich Defence unit. Clade would survive the move, and the war, with a total of 26 kills to his credit.

30

Bf 109E-7/Trop 'Black 8', flown by Leutnant

Werner Schroer, 8./JG 27, Ain El Gazala/Libya, April 1941

If Clade's G-6, above, signifies JG 27's retirement from the Mediterranean, then Schroer's E-7 perfectly illustrates its arrival some three years earlier, wearing as it does the original theatre camouflage scheme of sand (Sandgelb 79), with green (80) dapple over light blue (78) undersides. The upper surfaces were intended to blend in with the local terrain of sand, dotted with camelthorn bushes. Note too the individual numeral edged in the *Staffel* colour of red, and also the white tailwheel tyre; the latter was not an affectation, but a serious attempt (seen on many early arrivals in the desert) to reflect the sun's heat, and thus preserve the rubber of the tyre. Just four victory bars adorn Schroer's rudder here; there would, however, be many more to come.

31

Bf 109G-2/Trop 'Red 1', Hauptmann Werner Schroer, *Staffelkapitan* 8./JG 27, Rhodes, circa February 1943

After a brief stint with the *Geschwaderstab* (see profile 8), Schroer returned to 8.*Staffel* as its Kapitan in July 1942. Here, his G-2 does not carry the *Gruppe* badge, but does feature an early-style III.*Gruppe* wavy bar symbol aft of the fuselage cross. By now Schroer's score had risen to 60. After a further succession of postings, Major Schroer completed the war as Kommodore of JG 3 'Udet', having amassed a total of 114 victories, including 26 four-engined bombers.

32

Bf 109E-4 'Yellow 5', flown by Oberleutnant Erbo Graf von Kageneck, *Staffelkapitan* 9./JG 27, Sicily, May 1941

A *Staffelkapitan* (note the white metal pennant on the aerial mast) in Max Dobislav's III./JG 27 during the *Gruppe*'s brief sojourn on Sicily, von Kageneck's flew this suitably marked E-4, which displays the same mix of Balkan and Mediterranean markings as profile 27, albeit minus the *Gruppe* badge. He subsequently led his *Staffel* to Africa in December 1941. The last 2 of his 67 kills were achieved over the desert, before he himself was critically injured in an encounter with Hurricanes south of Agedabia on 24 December 1941. Although von Kageneck managed to return to base, he died of his wounds in a Naples hospital on 12 January 1942. At the time of his death he was JG 27's highest-scoring pilot.

33

Bf 109G-6/R6 'Red 13', flown by Feldwebel Heinrich Bartels, 11.JG/27, Kalamaki/Greece, circa September 1943

IV./JG 27 was activated in Greece in May 1943, operating in the eastern Mediterranean until transferred to Hungary in spring 1944. Bartel's G-6

'Kanonenboot' wears standard, freshly-applied, three-tone grey (74/75/76) camouflage, and also carries the *Gruppe*'s unique identification marking of two parellel bars (in the respective *Staffel* colours) aft of the fuselage cross. Note too the name *Marga* under the windscreen quarterlight, and Bartel's current tally of 56 kills on the rudder, plus the motif celebrating the award of the Knight's Cross for his 45th victory. His total score stood at 99 when he was reported missing in action after destroying a P-47 over Bonn on 23 December 1944. Bartels' remains were found strapped in the cockpit of his wrecked Bf 109G-10 ('Yellow 13' of 15./JG 27 – another *Marga*) in 1968.

34

Bf 109G-6 'Black Chevron/Triangle', flown by Hauptmann Karl Rammelt, *Gruppenkommandeur* II./JG 51, Tuscania/Italy, circa February 1944

Another famous *Jagdgeschwader* which contributed to the Luftwaffe's Bf 109 fighter presence in the Mediterranean was JG 51 'Moelders', whose II.*Gruppe* arrived in Tunisia in November 1942, and thereafter served in Sicily, Italy and the Balkans. Rammelt's G-6 is typical of this later period. It wears standard finish, displays JG 51's 'Buzzard's Head' badge and a *Gruppenkommandeur*'s 'triangle-in-a-chevron' combination, plus his personal emblem below the cockpit. It also illustrates II./JG 51's unusual practice of locating the *Gruppe* bar not aft of the fuselage cross, but ahead of both it and the aircraft's individual identity marking. Major Rammelt survived the war with 46 kills, although he was badly wounded whilst attacking B-24s of the Fifteenth AF over northern Italy in late 1944.

35

Bf 109G-2/Trop 'White 5', flown by Feldwebel Anton Hafner, 4./JG 51, Bizerta/Tunisia, November 1942

In contrast to the G-6 above, 'Toni' Hafner's G-2 sports the standard desert finish which most, if not all, of II./JG 51's aircraft were wearing upon first arriving in Tunisia. Note again the II.*Gruppe* bar ahead of the numeral. As his rudder indicates, Hafner had already scored 62 kills in Russia, to which he added a further 20 over the Mediterranean in short order. Subsequently commissioned, he later returned to the Eastern Front and III./JG 51. He was killed over East Prussia on 17 June 1944 dogfighting at low-level with a Yak-9. Despite being lost almost a year before Germany's final defeat, Hafner's astounding score of 204 victories remained unbeaten within JG 51.

36

Bf 109G-6/Trop 'White 12', flown by Oberfeldwebel Wilhelm Mink, 4./JG 51, Tuscania/Italy, circa January 1944

Another successful, and long-serving, NCO pilot of

II./JG 51, Mink was twice shot down over the English Channel in 1940, only to be rescued by a German U-boat on both occasions. With his score standing at 72, Wilhelm Mink's luck finally deserted him on 12 March 1945 when he fell victim to RAF fighters during a courier flight over Denmark in an unarmed Focke-Wulf Fw 58.

37
Bf 109F-4 'Black Chevron/Triangle and Bars', flown by Oberstleutnant Günther Freiherr von Maltzahn, *Geschwaderkommodore* JG 53, Comiso/Sicily, circa May 1942

Second of the three *Jagdgeschwader* which bore the brunt of the Mediterranean campaign, JG 53 'Pik-As' ('Ace of Spades') arrived in the area from the Eastern Front in December 1941. Kommodore from October 1940 to October 1943, von Maltzahn's flew more than a dozen Bf 109s whilst with JG 53, this F-4 being typically marked with a full set of kommodore's Stab markings, including a horizontal bar either side of the fuselage cross (that part of the bar aft of the cross should not be confused with the normal II.*Gruppe* symbol), which dated back to pre-war biplane days. The tip of the spinner is in the *Geschwaderstab* colour of blue. 'Henri' Maltzahn survived the war, adding another 13 kills to the 55 depicted here – all of these were achieved during his service with JG 53.

38
Bf 109G-6/R6 'Black Double Chevron', Major Jürgen Harder, *Gruppenkommandeur* I./JG 53, Maniago/Northern Italy, March 1944

A long-time member of 7./JG 53 (see profiles 47 and 48), Major Harder was promoted to the position of Kommandeur of I.*Gruppe* on 15 February 1944. His 'Kanonenboot', shown here, no longer displays the rudder victory bars which distinguished his earlier Bf 109. In January 1945 he became Kommodore of JG 11, but was killed the following month when he crashed near Berlin due to oxygen failure. His final score was 64.

39
Bf 109G-2/R1 Trop 'Yellow 13', flown by Leutnant Wilhelm Crinius, 3./JG 53, Bizerta/Tunisia, January 1943

Yellow 13' (Werk-Nr.10 804), with its rudder displaying 100 Eastern Front victories (for which Crinius received the Knight's Cross and Oak Leaves), plus 14 kills since scored over the Mediterranean, was prepared for Crinius' return to 3.*Staffel* after a brief posting to the *Gruppenstab* of I./JG 53. In the event, he failed to return from his interim appointment, being shot down by a US Spitfire off the Tunisian coast on 13 January 1943 and captured. The aircraft, retaining the markings and decorations shown here, was subsequently flown by Leutnant Gerhard Opel (ex-II./JG 26) as part of 2.*Staffel*.

40
Bf 109G-4 'Yellow 7', flown by Oberleutnant Wolfgant Tonne, *Staffelkapitan* 3./JG 53, Bizerta/Tunisia, February 1943

Crinius, had been wingman to *Staffelkapitan* Wolfgang Tonne. The latter's aircraft bore a similarly decorative rudder: Knight's Cross and Oak Leaves for 100 kills; a 101st Soviet victim, and then a string of victories over Tunisia. Tonne's G-4 displayed a segmented, rather than dappled, brown/green finish, with signs of overpainting around the cockpit and evidence of a previous identity beneath the 'Yellow 7'. With his score standing at 122, Tonne was killed when he crashed his badly shot-up 'Yellow 7' whilst attempting to land at Tunis-Protville on 20 April 1943.

41
Bf 109F-4 'White 1', flown by Oberleutnant Gerhard Michalski, *Staffelkapitan* 4./JG 53, Pantelleria, July 1942

The rudder of Michalski's green (74/75) dappled F-4 shows that he had scored 20 kills (including 14 over Russia) before arriving in the Mediterranean. There followed a succession of victories; 26 over Malta alone. By war's end he was Kommodore of JG 4, had achieved 73 kills – 13 of them USAAF bombers – and had himself been shot down six times! Ironically, he lost his life in a car crash nine months later.

42
Bf 109F-4/Z 'Black 1', flown by Hauptmann Kurt Brändle, *Staffelkapitan* 5./JG 53, Comiso/Sicily, circa February 1942

Although Brändle's distinctively-segmented F-4 displays no victory marks, he in fact scored 35 kills while serving with II./JG 53. He was to add another 145 after assuming command of II./JG 3 'Udet' in May 1942, only to be reported missing over the North Sea after an attack by heavily-escorted B-26 Marauders of the Ninth AF on Amsterdam-Schiphol airfield on 3 November 1943. His body was washed ashore some time later.

43
Bf 109F-4 'Black 2', flown by Oberfeldwebel Herbert Rollwage, 5./JG 53, Pantelleria, August 1942

Herbert Rollwage destroyed 71 aircraft during his time with II./JG 53, although one of the 12 Soviet kills which opened his score, as shown here, was subsequently disallowed. Among other victories yet to come would be 14 USAAF bombers, downed later in Defence of the Reich operations.

44
Bf 109G-6 'Black 2', flown by Oberfeldwebel Herbert Rollwage, 5./JG 53, Trapani/Sicily, July 1943

By July 1943 Rollwage was a *Schwarmführer* (section leader) within 5./JG 53; as witness the all-white rudder, upon which his first 11 Eastern Front victories were now correctly recorded. He was badly wounded shortly thereafter, but returned to become *Staffelkapitan* of 5./JG 53 in August 1944. He ended the war as Kommandeur of II./JG 106.

45
Bf 109G-6/R6 Trop 'Yellow 1', flown by Hauptmann Alfred Hammer, *Staffelkapitan* 6./JG 53, Cancello/Italy, August 1943

This lightly-dappled 'Kanonenboot', with distinct signs of overpainting behind the individual aircraft letter, was the first machine flown by Hammer during his long tenure as *Staffelkapitan* of 6./JG 53. On 8 January 1945 he took command of IV./JG 53, which he then led until the war's end. His final score totalled 26, and included 2 heavy bombers.

46
Bf 109G-6 'Black Double Chevron', flown by Major Franz Götz, *Gruppenkommandeur* III./JG 53, Orvieto/Italy, circa January 1944

Whether by virtue of his age (he celebrated his 31st birthday in January 1944), or by his unbroken length of frontline service (his first kill had been a Morane MS.406 on 14 May 1940), the Kommandeur of III./JG 53 was known as 'Altvater' ('Old Father') Götz, although not, one suspects, to his face! Shown here midway through his 27 months in command of the *Gruppe*, Götz's G-6 displays standard camouflage and markings of the period (note the fuselage cross is not black, but dark grey). Made Kommodore of JG 26 *Schlageter* on 28 January 1945, he ended the war with 63 kills.

47
Bf 109F-4/Z Trop 'White 5', flown by Leutnant Jürgen Harder, 7./JG 53, Martuba/Libya, June 1942

'White 5' wears standard desert camouflage and theatre markings, although the first 10 of the 17 victories shown here were, in fact, scored on the Eastern Front. Harder was one of three brothers, all fighter pilots, and all killed during the war. The name below the cockpit sill commemmorates Hauptmann Harro Harder, Kommandeur of III./JG 53, who had been shot down and killed off the Isle of Wight at the height of the Battle of Britain almost two years before.

48
Bf 109G-4/Trop 'White 1', Hauptmann Jürgen Harder, *Staffelkapitan* 7./JG 53, Trapani/Sicily, February 1943

Hauptmann Harder became *Staffelkapitan* of 7./JG 53 on 5 February 1943. Here, his G-4 bears a different camouflage scheme and 16 more kill markings, but the tribute to his brother remains, with a 'lucky' four-leaf clover below it. For Jürgen

Harder, however, luck was to run out on 17 February 1945 (see also profile 38).

49
Bf 109F-4 'White 2', flown by Leutnant Hermann Neuhoff, 7./JG 53, Comiso/Sicily, March 1942

Another 7.*Staffel* machine, Neuhoff's F-4 still wears the two-green (74/75) soft dapple which was standard prior to the unit's transfer to North Africa. The last 5 of the 36 kills displayed here were scored over the desert in December 1941. Deploying to Sicily for the attack on Malta, 4 more victories would be added before Neuhoff himself was accidentally shot down over Luqa on 10 April 1942 by a *Staffel* colleague who mistook him for a Hurricane – he was to spend the rest of the war as a PoW. It is debatable whether Leutnant Schöw, the offending party, included his *Schwarmführer* among the 15 kills he claimed prior to being listed missing in action over Stalingrad four months later – intriguingly, only one of these 15 victories was not scored in Russia!

50
Bf 109G-4/Z Trop 'Black 1', Oberleutnant Franz Schiess, *Staffelkapitan* 8./JG 53, Tunis-El Aouina/Tunisia, circa February 1943

Back to standard desert finish – albeit somewhat untidily oversprayed – for this 8.*Staffel* 'Kanonenboot'. The 38 victories depicted here had mostly been gained during Schiess' previous year's service as *Geschwader*-Adjutant. Could the overspray be concealing a set of Stab markings? Schiess added a further 29 kills during his six months at the head of 8./JG 53, before falling victim to P-38s off the Gulf of Naples on 2 September 1943.

51
Bf 109F-4/Z Trop 'Yellow 1', flown by Oberleutnant Franz Götz, *Staffelkapitan* 9./JG 53, Martuba/Libya, circa June 1942

A last look at the classic desert scheme of tan (79) over light blue (78), as worn by the F-4/Z Trop of Franz Götz prior to his promotion to the command of III.*Gruppe* (see profile 46).

52
Bf 109F-4 'Black Double Chevron', flown by Leutnant Heinz-Edgar Bär, *Gruppenkommandeur* I./JG 77, Southern Italy, July 1942

Last of the major triumvirate of *Jagdgeschwader* forming the Luftwaffe's fighter presence in the Mediterranean theatre, and by far the least known of the three, JG 77 did not arrive in Africa until one month before the Battle of El Alamein, the final turning point of the desert war. The *Geschwader's* lot thereafter was one of steady retreat – to Tunisia, Sicily and Italy – despite the best individual efforts of such outstanding aces as Heinz Bär, whose F-4 is shown here. Arguably the most seasoned *experte* of them all, he was in frontline service from the first weeks of the war to the very last – Bär opened his account with a French Hawk 75 on 25 September 1939 as a Feldwebel with I./JG 51, and finished with a P-47 over Bavaria on 28 April 1945 as an Oberstleutnant flying Me 262 jets with JV 44. No less than 218 Allied aircraft had fallen to 'Pritzl' Bär's guns in the intervening period.

53
Bf 109G-4/Trop 'Black Chevron', flown by Leutnant Heinz-Edgar Berres, *Gruppen*-Adjutant I./JG 77, Matmata/southern Tunisia, circa January 1943

Wearing a fresh coat of unfaded tan (79), with green (80) dapple overall, Berres' G-4 displays a standard *Gruppen*-Adjutant's chevron and, half-hidden by the wing, the red map of Great Britain within an ornate white capital 'L', which was the badge of I.(J)/LG 2, the *Gruppe*'s original identity prior to redesignation in January 1942. Berres subsequently became *Staffelkapitan* of 1./JG 77, claiming 53 victories before being shot down by British fighters while escorting a formation of Ju 52s across the Straits of Messina on 25 July 1943.

54
Bf 109G-6 'White 1', flown by Oberleutnant Ernst-Wilhelm Reinert, *Staffelkapitan* 1./JG 77, southern Italy, circa August 1943

Emerging as the top-scoring pilot of the Tunisian campaign (with 51 kills from January to April 1943), Reinert became Kapitan of I.*Staffel* in August 1943. His G-6 wears the *Geschwader* badge, introduced in April 1943, which gave the unit its name of JG 77 'Herz-As' ('Ace of Hearts'). By war's end Hauptmann Reinert was in charge of IV./JG 27, and had scored 174 aerial kills, with a further 16 aircraft destroyed on the ground.

55
Bf 109F-2/Top 'White 3', flown by Unteroffizier Horst Schlick, 1./JG 77, Bir El Abd/Egypt, November 1942

In a scheme typical of JG 77's early desert service, Werk-Nr.10 533 was discovered after the battle of El Alamein abandoned at Bir El Abd. The symbol aft of the fuselage cross is the badge of 1.*Staffel*, initially introduced after the Polish campaign, when the unit was still oerpating as I.(J)/LG 2, by one Harro Harder (see profiles 47 and 48). This marking actually dated back even further to Harder's days with the Legion Condor in Spain, where it was used as the emblem of his 1.J/88. The rudder victory bars (which some sources state as being black, rather than red) indicate two Eastern Front kills and six achieved since his arrival in the desert. Having escaped the El Alamein battlefield by truck, Schlick ended the war with a score of 36.

56
Bf 109E-4 'Black Chevron/Triangle', flown by Hauptmann Herbert Ihlefeld, *Gruppenkommandeur* I.(J)/LG 2, Molaoi/Greece, May 1941

Already an ace of the Legion Condor with seven kills to his credit, Ihlefeld joined I.(J)/LG 2 in 1938. He was Kommandeur from August 1940 until May 1942, by which time the *Gruppe* had been dedesignated I./JG 77. Shown here in Greece, his E-4 still wears the elaborate yellow Balkan campaign markings including rudder and tailplane, thin fuselage stripe, wingtips and even wing training edges. Indicative of the 'hand-me-down' aircraft from other units being flown by the *Gruppe* at this period, it still sports the badge of its previous owner, JG 52, below the windscreen quarterlight. After leaving JG 77, Ihlefeld served as Kommodore of five other *Geschwader*, before ending the war with a total of 130 victories.

57
Bf 109G-6/Trop 'Black 4', flown by Sottotenente Giuseppe Ruzzin 154a *Squadriglia*, 3° *Gruppo CT* (RA), Comiso/Sicily, July 1943

In addition to the Luftwaffe, four Italian *Gruppi* (plus some smaller units) flew the Bf 109 operationally in the Mediterranean theatre: two with the Regia Aeronautica (RA) prior to the Armistice of September 1943, and two with the Aeronautica Nazionale Repubblicana (ANR) alongside German forces in the closing months of the war. 'Black 4' is representative of machines taken straight over from the Luftwaffe, all of whose markings, with the exception of the white fuselage band, have been overpainted. The Italians have applied a stenciled individual aircraft number, the *squadriglia* identity on the white band, the 'Diavolo Rossi' ('Red Devil') *Gruppo* badge to the cowling and a plain white Savoia cross on the tail. Note that there are none of the insignia usually associated with Regia Aeronautica aircraft – the fasces symbols on fuselage and wing surfaces are absent, as too is the coat of arms normally centered on the tail cross. The RA strongly discouraged individualism, and preferred to credit victories to units as a whole, rather than to pilots.

58
Bf 109G-6/Trop 'White 7', flown by Tenente Ugo Drago, Comandante 363a *Squadriglia*, 150° *Gruppo CT* (RA), Sciacca/Sicily, May 1943

'White 7', the lucky number always carried by Italian ace Ugo Drago, displays slight differences from the 3° *Gruppo* aircraft above. Here, although all German markings have again been obliterated, stencilled fasces have been applied underwing and the *squadriglia* number has portrayed in the more usual manner. The 150° *Gruppo* 'Gigi Trei Osei' badge, however, has not been added.

59

Bf 109G-10/AS, flown by Maggiore Adriano Visconti, Comandante I° _Gruppo Caccia_ (ANR), Lonate Pozzolo (Varese)/Italy, February 1945

After the Italian Armistice, the two Axis Republican _Gruppi_ also received ex-Luftwaffe Bf 109s, as shown here by the sketchily-overpainted fuselage and tail markings (although the underwing crosses have been retained). The national insignia now consisted of the green-white-red tri-colour, usually fringed in yellow as shown. Note that Werk-Nr.491356 doesn't carry any individual number, but does wear the _Gruppo_'s 'Asso di Bastoni' ('Ace of Clubs') badge taken over from the RA's 153° _Gruppo CT_. Visconti claimed 7 kills with I° _Gruppo_ (to add to the 19 he scored with the RA). He was brutally murdered by Italian partisans soon after surrendering his _Gruppo_ to them on 29 April 1945.

60

Bf 109G-6 'Black 7', flown by Capitano Ugo Drago, Comandante 4a _Squadriglia_, II° _Gruppo Caccia_ (ANR), Aviano/Italy, November 1944

Although heavily-overpainted around the rear fuselage, Drago's G-6 retains both fuselage and wing crosses, as well as his lucky '7'. The stylish 'Gigi Trei Osei' badge on the nose, stencilled in white, is a simplified form of that previously carried by the RA's 150° _Gruppo_ CT when it fought alongside the various _Jagdwaffe Gruppen_ in Libya. It was designed and introduced in memory of Tenente Luigi 'Gigi' Caneppele, a pre-war Olympic glider pilot of some note, and combined his 3rd Class glider badge (the 'Three Birds' of the title) with a desert palm. Drago himself survived the war as the ANR's highest-scoring pilot with 11 kills to his credit.

BF 109 ACES – UNIFORM PLATES

1

Oberleutnant Hans-Joachim Marseille, depicted following the award of the Knight's Cross with Swords and Oak Leaves on 18 June 1942. He is wearing a tropical issue Luftwaffe sidecap and his distinctive, privately-purchased, leather jacket, with shoulder boards denoting rank. Marseille also favoured the standard-weight Luftwaffe flying suit trousers and boots for many sorties – during the extremely hot summer months he occasionally flew in shorts. A silk scarf prevented neck chafing.

2

Typical fighter pilot 'underess', well suited to the North African climate. An early bomber-style kapok-filled lifejacket was worn over a khaki shirt, with the popular British-style pith helmet, boasting a Luftwaffe eagle insignia sewn on a triangular 'tropical' backing, protecting the pilot's head from the sun. Tinted one-piece goggles are pushed up over the helmet, the latter being the only item shed by the pilot prior to embarking on a sortie.

3

A leutnant, circa early 1942, wearing a sheepskin-lined suede jacket with rank epaulettes and leather gloves. The typically baggy Luftwaffe tropical issue trousers were ideal for coping with the daytime heat, as well as offering some protection against the cold nights. The sidearm would have been a Luger or Walther P38, and regulation leather/canvas combination shoes are worn. A white top (in this case very dirty!) to the standard service cap denoted 'summer' issue.

4

Major Joachim Müncheberg wears a 'Mae West' inflatable lifejacket over his flying overalls. The Knight's Cross is worn at the throat and the rank pennant is sewn on the sleeve. The uniform trousers were sometimes worn loose as shown, rather than tucked into flying boots, and were made roomy enough to contain a sidearm, knife and maps. A tropical issue cap is worn, and among the lifejacket attachments is an oxygen mask lead and map reading torch.

5

An oberleutnant wearing a typical mix of Luftwaffe and Afrika Korps kit, the headgear being that issued to ground forces and customised with the addition of a Luftwaffe patch. The short jerkin, or 'Fliegerblouse', with elasticated waistband, has a rank pennant, and the wearer is a Knight's Cross holder. Tropical flying trousers are tucked into leather flying boots, with the left boot topped by a rubber 'bandolier' of signal flare cartridges.

6

Oberfeldwebel Otto Schulz of II./JG 27 in late 1941. Standard officer's sidecap is worn, with the Knight's Cross at the throat and the two 'pips' of rank on the shoulder boards. A mission clasp is pinned above the left breast pocket, with the Iron Cross 1st Class just below it. The pilot proudly wears his wings (left) and wound badge just above his belt, whilst the regulation cloth Luftwaffe eagle patch has been sewn onto his crumpled tunic above the right breast pocket.